Monitoring and Evaluation
of
Agriculture and Rural Development
Projects

Dennis J. Casley and Denis A. Lury

Published for The World Bank
THE JOHNS HOPKINS UNIVERSITY PRESS
Baltimore & London

The Johns Hopkins University Press
Baltimore, Maryland 21218, U.S.A.

Library of Congress Cataloging in Publication Data

Casley, D. J.
  Monitoring and evaluation of agriculture and rural development projects.

  Includes index.
    1. Agricultural development projects—Management.   2. Rural development proj-
jects—Management.   I. Lury, D. A.   II. World Bank.   III. Title.
HD1415.C33   351.82'33   82-7126
ISBN 0-8018-2910-0 (pbk.)   AACR2

PREFACE

Awareness of the need to monitor and evaluate agriculture and rural development projects during implementation has developed rapidly in recent years; a large proportion of projects include specific components for this purpose.

In many instances such awareness has not been matched by the formulation of systems that enable monitoring and evaluation to be carried out effectively at reasonable cost. The Monitoring and Evaluation Unit under the Agriculture and Rural Development Department in the World Bank has, in the last three years, conducted analytical research, regional workshops and case studies of experiences with monitoring and evaluation systems.

This Handbook, prepared in conjunction with a set of Guidelines, sets out in some detail the issues pertinent to the implementation of monitoring and evaluation systems and discusses various aspects of data collection and analysis required for them. The initial preparatory work was a joint endeavour of the staff of the Monitoring and Evaluation Unit. In the preparation of the Handbook the authors synthesized this work and utilized their own special expertise in these fields.

As indicated in the Introduction it is hoped that this Handbook will be of assistance to those who are responsible for implementing monitoring and evaluation systems, as well as to project management. The Handbook is not a comprehensive textbook. Rather, it elaborates on concepts deemed essential for monitoring and evaluating projects during implementation and provides a framework for such systems. The basic purpose of monitoring and evaluation systems is to provide a tool for project management; in discussing data collection and analysis issues the Handbook emphasizes pragmatic and simpler approaches that are sometimes neglected in this advanced age of computers and sophisticated econometric analyses.

Leif E. Christoffersen
Assistant Director
Agriculture and Rural Development Department

CONTENTS

v

CONTENTS

CONTENTS

Figures Listed

Tables Listed

CONTENTS

PART 1:
THE FRAMEWORK FOR MONITORING AND EVALUATION OF AGRICULTURE AND RURAL DEVELOPMENT PROJECTS

1.1 Introduction

This Handbook has been prepared in conjunction with Guidelines for the Design of Monitoring and Evaluation Systems for Agriculture and Rural Development Projects. In the Guidelines, issues of significance to the establishment and operation of such systems are introduced without elaboration: this Handbook provides further discussion of these issues and sets out recommendations for dealing with them.

The Handbook is directed primarily to those working within a specific project, whose task it is to operate the monitoring and/or evaluation systems, and secondly to those responsible for management of the project so that they may better assess what they can expect their monitoring and evaluation operations to provide.

With this as the aim, one constraint becomes immediately apparent; while the correct structure of the project management team, and its relationship with the monitoring and evaluation staff is of prime importance in most instances, those using the Handbook will be faced with a situation where these issues have already been decided. Some adjustment may be possible as the project gets under way, but it may be difficult either to alter the management structure or to make a substantial realignment of the monitoring and evaluation staff within that structure.

Another major concern is the role of evaluation in assessing management performance. Once again, a Handbook addressing evaluation in a project specific context can offer little discussion of this for the assessment of individual performance is carried out elsewhere; and comparisons across a set of projects or assessments of projects against the national development programs will be part of a central evaluation effort that may or may not include responsibilities for individual project evaluations.

Moreover, recommendations for the collection of baseline data *before* project implementation are better directed to those responsible for the preparation of projects, than to those whose first contact with the project is when implementation is beginning.

These issues cannot be entirely ignored, since activities have to be considered in their overall framework. However, they are often addressed in a warning context. Given a particular management structure and location of the monitoring and evaluation staff, the objectives of the latter must be set accordingly. In particular, attempts to remedy the lack of a pre-project data base in the early days must not absorb such a proportion of the resources available that the current information needs of management are neglected.

Because it is oriented towards a wide readership, the Handbook includes some elementary (but fundamental) material. The experienced practitioner may find some sections redundant or repetitive, but it is hoped that the Handbook will be "consulted" as well as "read".

As the title indicates, the Handbook addresses agriculture and rural projects. These encompass a wide range of project types, but there are common features for which general monitoring and evaluation principles may be developed. These are

(a) The target population is the rural population (or part of it) within a specific area of a country;

(b) The impact is generally to be achieved by providing an impetus to change in farming patterns or farmers' response to perceived constraints, so as to alleviate these constraints. (Other components may be included with this main focus, but rarely in isolation of it);

(c) The majority of the intended beneficiaries are usually drawn from the small farmers of the area (by definition in the case of rural development projects, but in relative terms this applies to many agriculture projects also), but general change at the community level is also commonly envisaged;

(d) Although the technological package supplied may be well proven—both experimentally and in terms of experience—in other areas, the response of the recipients in a particular case cannot always be predicted with precision in advance of project implementation; and

(e) The intended impact may be obscured in the short term by natural and man-made phenomena outside project management control, which may have larger effects than the project.

Viewing the project objectives as a sequence, shown below, will assist in distinguishing between the functions of monitoring and evaluation.

(a) The immediate objective is to provide *inputs* that are necessary to achieve agriculture and/or rural development.

2

Example: Inputs may be provided in the form of a physical facility such as an irrigation system or health clinic, advice which the beneficiaries are to be encouraged to adopt, or supply of such services as credit, fertilizer, water supply, medicine, etc.

(b) It is expected that the use of these inputs will result in *outputs by* the project beneficiaries.

Example: In many projects the outputs are in terms of crop or livestock production but outputs may also be skill acquisition, larger school attendance, greater use of health facilities, etc.

(c) These outputs will, in turn, generate *effects* amongst the target population.

Example: Commonly, the effect will be a change in income and expenditure levels and patterns but health and other social consequences may also be expected.

(d) Finally, these effects will have an *impact* on the social and economic life of the community.

Example: As a result of improved incomes, services may develop in the area providing wider income and employment opportunities, or, as a result of better health and education, the general quality of life may improve.

Effects and impact shade into each other; the difference is largely one of degree along three dimensions; namely, time, scale, and scope. Effects will show through sooner, apply to the direct beneficiaries and relate to specific aspects of rural activity. The impact measures the final total result, taking into account direct and indirect effects and allowing for diffusion and imitation that produce changes in the community as a whole.

Physical inputs and outputs are measurable, at least in theory. The effects and impact, however, are not only difficult to measure but may not easily be attributed to the stimulus of the project. The project design will have adopted certain 'reasonable' assumptions regarding the likely effects and their causal links to the project, but these require constant review in the light of developments during actual project implementation.

Monitoring and evaluation systems must be designed to reflect the achievement of these project objectives as expressed in targets to be met over time. Imprecise targets to be achieved at indeterminate times will lead to loosely-defined monitoring. However, a balance must be maintained; precision must reflect what is possible. There is a danger that excessive precision of targets and timing may lead to a excessively large-scale data collection effort. This effort will not usually be cost effective in terms of its contribution to decision-taking.

The costs of the monitoring and evaluation effort must be related to the likely returns in improving implementation and the spin-off effects in improving the design of follow-up or similar projects.

1.2 Monitoring and Evaluation: The Basic Framework

The monitoring and evaluation functions are related but distinct. Monitoring is the provision of information, and the use of that information, to enable management to assess progress of implementation and take timely decisions to ensure that progress is maintained according to schedule. Monitoring assesses whether project *inputs* are being delivered, are being used as intended, and are having the *initial effects* as planned. Monitoring is an internal project activity, an essential part of good management practice and therefore an integral part of day-to-day management. *Evaluation* assesses the overall project *effects*, both intentional and unintentional, and their *impact*. It involves comparisons requiring information from outside the project either in time, area, or population. The relative roles of monitoring and evaluation will vary with the type of project. The supply over a wide area of a well-proven package aimed at a specific crop or farm activity needs careful monitoring, but possibly less emphasis on evaluation. An innovative, but small-scale, project may be easier to monitor, but evaluation will be both difficult and critical.

The initial steps for designing monitoring and evaluation systems are
(a) A review of the project objectives in order to systematize them according to the sequence indicated in Section 1.1 above; and
(b) Identification of the users of both the monitoring and evaluation information. For monitoring, the users will be the hierarchy of project management. The type of information transmittal will be geared to the needs of each level of project management. The users of evaluation analysis range from project management through the responsible directorate/ministry, to the national planners.

Evaluation will draw on the data generated by the monitoring system to help explain the trends in effects and impact of the project. Monitoring data may reveal significant departure from expectations which may warrant the undertaking of an *on-going evaluation* exercise to examine the assumptions and premises on which the project design was based. Such a review, as also in the case of ex-post evaluation, can be of great value to sectoral management in its policy formulation role.

Monitoring must be integrated within the project management structure; but evaluation, with its wider horizons requiring comparative information, is not necessarily such an integral component. A central evaluation facility may be justified on the grounds that:
(a) The demanding professional skills required to interpret evaluation data are either unavailable or uneconomic for each project individually;
(b) The data needed extend from before a project is initiated to a period long past its completion; and

4

(c) It is necessary to have comparable data from within and without the project population to achieve some form of rudimentary control comparisons.

Although the design and analytical facility for evaluation may be centralized, the data collection resources within a project will be used to provide much of the required data. If the same unit is collecting data both for eventual evaluation and for quick, timely monitoring, the latter must not suffer due to the greater demands of the former.

1.3 Monitoring and Management

Within the project the primary need is to monitor progress. Given that the project has been carefully appraised; i.e., that there is a strong *a priori* assumption that certain stimuli and inputs will achieve certain outputs, effects and impact, the role of management in the early implementation phase is to create the conditions that will allow this chain of events to occur.

In the early years of project implementation the emphasis will be on monitoring the physical and financial progress and the delivery of the inputs to the intended recipients. The main source for this aspect of monitoring is properly organized project records. The other concern of management, at this stage, is the use to which these inputs are put and the reaction of the recipients.

Adoption rates, and more important, repeat adoption rates, give management a strong inference whether the project is succeeding or not. Information on the recipients' attitudes and perceptions is important in order to explain any departure in response behavior to that postulated in the project design. Such unpredicted behavior may determine the success or failure of the project.

The information required for monitoring of project implementation does not require complex data systems. A monitoring system will exist even if it is merely a subjective accumulation of impressions by project staff. If common sense rules of good standard management practice are adhered to, the monitoring system can be limited to the minimum of parameters to be recorded regularly over time. The goal is to make the data collection as objective as possible, and to ensure, above all, that the means exist for fast collation, summarization and presentation of the information to the decision-makers. The need for recording of progress by project staff together with their subjective impressions is not replaced by the existence of a monitoring unit—rather, such reporting is an essential source for the system.

Once management has satisfied itself that the delivery system is working, its attention should shift to the outputs generated; i.e., are they materializing according to expectation? Focus on output measurements must not, however, be at the expense of monitoring the input delivery system. The measurement of outputs is more properly a function of evaluation, for identifying trends is not an easy task in view of the exogenous influences at work, and is often impossible without an extended time-series over many years. Nevertheless, management will need estimates of production, and the means of obtaining them is discussed later in this Handbook.

The key to successful monitoring is the provision of regular, timely, decision-oriented information to the project management. This can be achieved if the necessary staff are in place early, are seen to be part of the management team, and are given guidance on the priority information needs of management.

1.4 Evaluation: An Assessment of Results

Evaluation aims to determine whether the project objectives set in terms of expected outputs, effects and impact are being, or will be, met. This leads to an assessment of the results achieved, and the lessons to be drawn for future improvements in a later phase or in similar projects elsewhere.

Output levels are a measure of the result of the input utilization by the beneficiaries. If the changes in outputs are considerable, they may be detected even during the implementation phase of a project. In other cases, the effects, for example, on health, arising out of the provision of health services as an input, may not be quick to appear. And the *impact* of such effects on the general quality of life of the community will, in most cases, be a slow-developing process. An evaluation system will require the development of a series of data commencing before the project is implemented and continuing well past the completion of the implementation period. Unlike a monitoring system with its emphasis on rapid assessment, an evaluation system requires a longer time span before even tentative conclusions can be drawn.

The extent to which even moderate variability can affect the picture is not always appreciated. The following example shows how it might operate, by simulating three sets of results obtained by subjecting an underlying simple linear trend to random variation. The trend is from 1,000 to 2,000 over five years: the variability used in the simulations is normal (in the statistical sense) with a standard deviation of 15 percent of the trend values. The three series were the first obtained; there has been no searching for extreme patterns.

Table 1: Random Variations About a Trend

Year	Trend	Simulations		
		1	2	3
0	1000	950	632	1163
1	1200	1170	1284	1423
2	1400	1213	1522	1316
3	1600	1596	1714	1748
4	1800	1942	1945	1498
5	2000	1658	2192	1668

The simulations are equally likely, and all assume that the underlying trend is in fact taking place as the basis on which random variation is imposed. Suppose, however, that three project areas all had the trend figures as annual targets; and that they produced the series shown above as their results. Would they be regarded as equally successful? It is important, therefore, that the role of natural variability should be considered before any hasty conclusions about relative performance are drawn.

Evaluation will go beyond quantifying changes that occur to an assessment of the contribution of the project in achieving them. But the project area and the beneficiaries are subject also to exogenous influences beyond the control of the project; these are unpredictable in their timing and magnitude. Therefore, establishment of causality between project inputs and the effects and impact poses the most difficult problem in evaluation. These issues are taken up in Part 3.

In addition to the analysis of data series over time, evaluation will usually require in-depth studies of the validity of certain assumptions implicit in the project justification. Are the socioeconomic determinants of intended beneficiary behavior understood? Will the market for the intended incremental production hold stable when the increased supply enters the system? Will the income generated benefit the dependents of the beneficiary or be siphoned away from the household? These are questions that project management, taxed with many implementation issues, has no time to consider unless the unexpected impact of one or more of these factors threatens the implementation process; and by then it may be too late. Such studies will contribute to the process of ongoing evaluation referred to earlier and some involvement by the monitoring team will be required. This illustrates the point, returned to below, and in Parts 2 and 3, that there are close relationships, in practice, between monitoring and evaluation despite their distinct functions.

To what extent should data collected within a project be standardized to contribute to inter-project comparisons or national data series? Viewed as a single project covering only a small area of the country and unique in the type of development to be undertaken, the answer may be very little. But agriculture and rural development projects often cover large geographical areas individually, and collectively cover a large part of the country. Monitoring and evaluation resources within each project may, in total, be a very significant part of the total statistical resource of the country. There is then a strong case for these resources to contribute to the broader aspects of information needs, by introducing standardization of survey design, data content, or analytical treatment to the extent appropriate. Many of the data needs on outputs and socioeconomic aspects of the population are similar in different projects and similar again to national requirements. Such standardization need not divert attention from the needs of project management—indeed may help to assist project management identify its broader needs.

Moreover, if confined solely within a project, the data collection and analytical skills may be dissipated, once the project is completed. This would not benefit the overall state of statistical development in the country. The mutually supportive roles of national data agencies and project monitoring and evaluation units need to be identified and exploited for the benefit of both.

1.5   Monitoring and Evaluation in the Project Cycle

In Bank terminology, the Project Cycle is:

1. Identification
2. Preparation
3. Appraisal
4. Implementation, and
5. Completion

Figure 1 sets out these stages and identifies the main data needs and their possible sources of supply. The summary phases in the Figure cannot of course cover many of the aspects that arise across the whole project range, but they give the broad picture.

The first three phases precede the actual project activities in the field, but they are shown here to indicate the background. The information will also help those staff in ongoing projects who are, as is often the case, involved in preparing proposals for further phases of project activities, either by continuation or by repetition elsewhere. Their presence also makes clear that some monitoring and evaluation activities in phases 4 and 5 can hope to succeed only if they are planned and provided for in the first three phases.

Broadly speaking, monitoring is in phase 4 and evaluation is in phase 5. But, as already indicated, and discussed further in Parts 2 and 3, there is no hard line between the two functions in practice, of the kind drawn across Figure 1 to illustrate the separate phases.

The essential decision in the pre-implementation phases is whether (or not) to conduct a major baseline survey. The decision on this affects the whole evaluation effort during future years. A baseline survey using a probability sample is justified if it is intended to collect the same information at other points during the project cycle, including the period after project disbursement is completed. The emphasis for evaluation will be on a longitudinal comparison of these probability samples. Desirable though it may be for such a survey to be mounted, executed and analyzed in an appropriate pre-implementation phase, it requires the existence of a capable survey organization on hand. Contracting it to a university institute with experience and capability only in the area of micro in-depth studies will not usually be desirable. If no suitable organization exists, the alternative is to rely on the monitoring system for simpler, but more continuous longitudinal indicators supplemented by small-scale, possibly purposively selected, sample surveys that aim to reflect change in relative terms across groups (including the concept of at least a quasi-control group). For this the involvement of local institutions may well prove helpful.

Some data on the demographic and socioeconomic characteristics of the population are necessary at the project preparation stage, quite apart from any baseline survey. A quick interview survey using a purposive or quota sample, possibly including a community-level questionnaire, may well serve this need. Only approximate levels are required at this stage; thus the issues of recall periods and oversimplification of the questions, inevitable in a short interview, may not be limiting. The cost of such a survey will be modest relative to other project preparation and appraisal costs, but the reason why it is rarely accomplished may be the lack of a survey capability in the country.

A socioeconomic investigation by one or two skilled professionals into the attitudes and constraints affecting the likely beneficiaries could also be of great value at this stage. Many projects assume certain beneficiary responses which do not materialize due to a lack of insight into these.

Once the implementation stage is reached the monitoring system remains within its objectives. It is not the role of this system to undertake, late in the cycle, surveys that should have been undertaken earlier.

Some evaluation time-series data will be available at the completion stage only if there has been a baseline survey. If this was accomplished it will need to be repeated twice or thrice over the years. This will be a substantial commitment, and if the same external agency that executed the baseline can be contracted for the further rounds, so much the better. Not only is continuity achieved but the internal monitoring system is not diverted into these time-consuming exercises.

Each project—with the possible exception of straightforward follow-up efforts—is to some extent unique. A typology and timetable for monitoring and evaluation would, therefore, be misleading in masking the individuality of a project that requires a 'tailor-made' information system.

Part 4 discusses the choice of indicators in some detail and the sources of the data are discussed in Parts 5, 6, and 7. Before proceeding to these issues the distinctive features of monitoring and evaluation are considered further in Parts 2 and 3.

Figure 1: Project Cycle: Data Requirements and Sources

Stage	Data required	Source
Identification	Project area, current outputs, beneficiaries, & environment.	Administrative files, census, national and local survey, MOA estimates.
Preparation	Technical inputs.	Professional review. Feasibility studies.
	Socio-economic characteristics of project population.	Existing rural surveys and/or *specially commissioned* ad hoc study by NSO or others.
	Market information.	Administrative files, existing market survey, and/or *specially commissioned* ad hoc market study.
	Attitudes of beneficiaries, constraints affecting them.	*Specially commissioned* case study.
	If base-line survey required for long term evaluation, specify survey content, basic approach, and scale.	*Specially commissioned* baseline survey (preferably by NSO).
Appraisal	Financial and O.&M.	Ministries, district organizations, agencies.
	Macroeconomic factors.	National accounts, sector studies, development plans.
	Project area indicators.	Area specific data generally available and results obtained at earlier stages.
Implementation Monitoring	(a) Financial Staff Construction Physical inputs Service inputs Marketing Input usage Production Adoption rates	Much of these data will come from internal project activity (see Part 2), either as part of regular reporting process or as the result of special enquiries mounted from project resources.
	(b) Beneficiary reaction Problem identification *If previously planned,* Mid-implementation Project survey Time-series and external data	Here the borderline between monitoring and evaluation is blurred. Case studies and surveys, and detailed analyses, may require resources from outside the project.
Completion Evaluation	(a) *If previously planned* Post-project survey. Coordination of data for review and analysis.	See entry directly above. Evaluation will require analysis involving persons outside project.
	(b) Long-term evaluation of persistence of results.	*Specially commissioned* survey and/or case studies. Relevant results from continuing NSO & MOA activities.

PART 2:
MONITORING

2.1   Managing and Monitoring

The interactions between monitoring and evaluation have been outlined in Part 1. The existence of these interactions has led to a common perception that monitoring and evaluation are inseparable, as indicated by the ubiquitous abbreviation, M&E. But there are clear distinctions in their roles as explained in Part 1; and experience shows that monitoring is not successful unless distinguished from evaluation. Only then will it play its proper role as a management tool within the project's management information system.

Monitoring arrangements must be regarded as part of general management, and be estimated for and funded accordingly. A common complaint by managers is that monitoring units do not provide information that is relevant or timely. This is usually the case where the unit is imposed on management as a parallel organization, leading managers to conclude that its main function is to provide information to outsiders, with little or no commitment to the project. This will not happen if monitoring is viewed as a management function, under the control of managers, designed to meet their information needs, and working to an agreed timetable, with given resources.

In general, independent consultants should be considered only for the following monitoring functions:

(a) advice on general design of the monitoring system at project preparation stage;

(b) specific ad hoc enquiries, when they are brought in with precise terms of reference as to what they do, to whom they are responsible, and how they report. These could include technological/scientific, quality control, and causal studies; and

(c) training and advice on survey methodology.

Evaluation is a different matter, and must include participants from outside the project management: in this context, evaluation of the project results and of project management performance includes evaluation of its monitoring performance.

This separation of monitoring and evaluation reduces, what a Workshop identified as, "the fear of the evaluation components of the M&E unit . . . which infects all divisions of the project. In these circumstances the monitoring unit can make a very limited contribution to the project's progress." The integration of monitoring into the management activity will not of course remove all the suspicions that may be felt against any group who carry out this type of function.

This emphasis on management monitoring does not require management to identify all its information needs from the outset. Such identification will be a continuous process as implementation proceeds, involving interaction and advice flowing upwards as well as downwards. But it must be unmistakably clear whose needs have priority and who has the final decision.

Difficult institutional problems may occur with this manager-oriented approach to monitoring in complex projects involving several different agencies, as commonly encountered in agriculture and rural development projects. But these difficulties do not relate to monitoring alone: they are basic to the setting up of an effective management structure for the project. If management responsibilities are clearly defined, it is not difficult to fit in the monitoring unit appropriately and to identify the reporting channels to the separate lines of management. If an effective management structure cannot be set up there is little point in proceeding at all. So the argument is the same as before: the managerial function includes monitoring what it is doing; management can only be effective when it has feedback.

When several specialized agencies are each responsible for separate components of a project, the 'top' management may be primarily a coordinating body. In such a case, the design of the project should have considered not only the location of a central monitoring unit (ensuring a flow of the required information to the coordinating management), but also the arrangements for the exchange of information between the line agencies involved in the various levels of field operations. A responsibility of the monitoring unit then is to develop these lines of communication so that the needs of each management level are met and unnecessary duplication in information systems is avoided.

Organized in this way, monitoring will still produce information which needs to be carried forward for evaluation. An example is given by a project monitoring report that included a section "on the individual performance of settlers, in which a revealing feature has been that the weaker ones are mostly the local people . . . One reason for this appears to be that their standards are limited to what was on offer before the project began . . . Comparatively, those who have come from outside seem to work harder and have a greater desire to succeed." Arrangements must be made to ensure that monitoring results of this kind are available to those engaged in evaluation: indeed, evaluators should have access to all monitoring data. This should not cause difficulties if management has provided itself with a satisfactory monitoring system and acted on its information during project implementation. In these circumstances, they will want evaluation to take this into account.

Sometimes monitoring may require close integration with evaluation, because it is expected that monitoring data is of particular importance in causal explanation (see Part 3). Special care is needed then to prepare and coordinate the scope and timing of data collection.

2.2 Specification of Objectives and Targets

Detailed specification of project objectives facilitates monitoring as the following cases show.

CASE A   The project involves completion of a settlement scheme and the improvement of drainage on low-lying coastal lands. In the settlement part, 8,000 acres of rolling land will be cleared for sugar cane and settled by 800 families. A sugar tramline extension, road improvements, and sites and services for a new township are to be provided. The drainage improvement involves reconstruction of seawalls and of internal drains on 26,400 acres of sugar cane land. There are physical production targets of sugar cane.

This project is relatively straightforward. Its operations can be functionally divided and monitored accordingly, by reports of money spent, work completed, etc., flowing to a small unit under the project manager. Since the crop is centrally processed and paid for, production, yields, and crop income can be recorded at the centre and made available with little trouble.

CASE B   Major objectives are:

(a) to extend multiple cropping technology coupled with improved livestock activities to increase the intensity of land use as well as raise farm incomes;
(b) to provide additional employment opportunities for subsistence farmers and in particular, landless labourers;
(c) to support research activities in multiple cropping; and
(d) to improve rural welfare through health support services.

Secondary objectives include infrastructural improvements. Here the objectives are more numerous, and multi-sectoral in nature. They will require evaluation, but their description gives no obvious guidance for the monitoring system. However, implementation targets that will lead to meeting these objectives will normally have been specified in the project documentation. The task of the monitoring unit in consultation with management will be to choose the indicators for measuring these target levels and the units in which they are to be measured. Questions that arise include how to measure land use intensity, incomes,

13

employment creation, rural welfare, etc., and the relative weights to be given each of these aspects (including which of them can be tackled only through longitudinal evaluation surveys).

Thus, for monitoring, it is the operational *targets* shown as the intermediate steps to reaching the project objectives that are particularly relevant. The danger of vague objectives is that the target specifications may also be vague or inappropriate. It is the job of senior management to break down the specified targets to the area or functional level for use by the lower echelons of management, and the job of monitoring is to measure progress against these target levels.

The information obtained in the monitoring process will sometimes lead to adjustments in the targets for various project components and the resources allocated to meet them. This, in turn, will cause adjustments in the monitoring system, and so on.

Targets for financial disbursements, staffing ratios, and infrastructure construction should be specified with considerable precision and monitored with accuracy. Beneficiary reactions and outputs can be targeted, but excessive precision in stating these should be avoided. The difficulty of monitoring output targets when random exogenous influences are involved was demonstrated in the hypothetical example of trend increases in Part 1. Year-to-year figures may give little guidance until a lengthy time-series has been established and the examination of such time-series is more properly an evaluation task.

2.3 Monitoring and Financial Accounting

Management accounting procedures and financial reports play a major part in most information systems for management control: their relationship to other monitoring information must be recognized. Financial controllers need to be involved by project management in the design of the monitoring system. They and their staff need to be involved throughout both in financial and nonfinancial monitoring. Plans for financial monitoring also need to be thoroughly examined by management to see that the financial reporting system will—like the rest of the monitoring arrangements—provide the needed data when it is wanted. Accountancy units are sometimes as poor as other monitoring units in providing timely, well-presented data.

Consistency of classification across the accounts produced for planning, reporting, and for comparison with monitoring data is essential. In particular, the distinction between capital and current expenditure has to be applied carefully. An example, at a disaggregated level, is the need for expenditure on small tools to be recorded consistently under the same head of expenditure. Consistency of valuation is also essential, or an agreed system of adjustments must be followed.

Accounting information must not be left passive or in isolation. The comparison of statements of expenditures and of physical progress helps to prevent

14

cost overruns, and improves control over worksite performance. It also acts as a check. One project, for example, states there was a tendency for some sectoral reports from agencies and ministries to exaggerate their performance; but that if financial accounts did not show that the relevant amount had been disbursed, the physical performance report was ignored until otherwise confirmed.

The financial report is often, but not necessarily always, the more accurate of the two. Sometimes the financial information is delayed: when, for instance, small-scale construction schemes are carried out locally by several agencies, the information is often received quite some time after the physical work is completed. The financial statements may even be incorrect. In a project involving freezing and canning plants, both profit margins and sales figures were understated and production costs and overheads overstated for the purpose of tax evasion. Only vigilant, on-the-spot management supported by cross comparisons can prevent these offences. Once again, financial and physical progress monitoring must work hand-in-hand.

2.4   Monitoring of Staffing

The first step in monitoring project staffing is to check from administrative records whether the recruitment of project staff is proceeding according to the timetable. If specialist staff are not being recruited on time, or there have been delays in setting up offices in one or more areas, urgent action must be taken, and the timetables re-programmed. Managers should, of course, have already identified crucial elements and kept a watch on them. Again, a flexible, iterative system is required, for what is crucial at one stage may differ from what is crucial at another.

Staff turnover is a major and persistent problem. It is particularly serious for staff in key positions, since they are usually more prone to offers of attractive alternative employment. For example, if the full potential of computers is not always realized, it is because it is difficult to keep competent analysts in their posts long enough to see large scale processing through to completion. Recommendations on the design and analysis of inquiries have been framed in this Handbook with this constraint in mind. Close monitoring of staff turnover will assist in the development of contingency arrangements and the optimal deployment of the available resources.

Monitoring the quality of work undertaken by the staff is more difficult. In an extension project, for example, the intended buildup of the numbers of extension agents will be clearly specified and can be monitored with relative ease. The delivery of the extension service can also be monitored by recording the time spent in the field and the number of visits made. In most cases, these indicators will provide a reasonable guide to the potential efficiency of the service and will also enable a first ranking to be made of the levels of performance. For this purpose, extension agents must keep individual logs of their activities, and the supervisors of the extension services must use a standard reporting format. Although, of necessity, subjective in nature, the assessment of performance against common criteria will enable preliminary judgements to be made. If the criteria can be agreed upon, a simple scale for measuring performance may be introduced. An example is shown in Figure 2 in which assessment on a scale of five levels is required.

Figure 2: Assessment of Extension Agents Performance

CRITERIA	Scale				
	1	2	3	4	5
A. Proportion of time in field					
B. Adherence to timetable					
C. Knowledge of techniques					
D. Performance at observed sessions					
E. Farmers' feedback					
F. Initiative					

It will be difficult at the beginning to get a large number of supervisors to operate these scales uniformly, and meaningfully. A universal rating of average will not help. An even number of points in the scale helps to prevent clustering on the mid-point. The distribution of ratings should itself be monitored to ensure that the basic purpose of detecting exceptional performances (both good and bad), is achieved. Improved standardization of assessment should be possible as the project proceeds and comparisons are made with the more direct indicators.

Other measures of extension effort can be devised; since the main concern is to encourage change in farming patterns and behavior, the ultimate monitoring indicators are the adoption rates for the recommendations made and their derivatives, e.g., farmer demand for inputs. These can be monitored using a sample survey of recipients of the service; their reaction to the service provided can also be included. This measurement of adoption as a monitoring tool is taken up in some detail in Part 4. However, the results of these indicators are not the final answer. At a Bank workshop, it was reported that, "Some extension workers feel that the present performance indicator, farm productivity, serves to discourage those assigned in depressed or poor communities where farm production inputs are rarely available and/or farmers cannot afford to use them. They suggest that additional indicators apart from productivity be devised to measure their individual performance." The staff reporting system outlined in the preceding paragraph will go some way to meeting this request. In addition, the reports will throw light on the supervisors' capabilities. This quote also illustrates the importance of building morale and team spirit. The monitoring system should take account of this also, by detecting evidence of changes in these attributes.

2.5 Monitoring of Infrastructure and Delivery Systems

Progress of physical construction according to a predetermined critical path and measurement of input delivery against targets are obvious monitoring tasks. The choice of indicators is usually clearcut.

Special checks on the quality of construction and the materials used for it are required. These should be planned and carried out by those technically competent. Some checks will be timed to coincide with stages of completion, and are often specified in the contracts. Inspections at random intervals help to ensure that substandard materials are not being substituted, or that work is not being skimped. Identification of bad work even after completion is better than uncritical acceptance of whatever has been done. One of the most effective ways of preventing things going wrong is to make it evident that continuous monitoring is going on; and that evidence of deficiencies is being promptly transmitted to the management for remedial action.

Monitoring in this context is a wider function than that carried out by a small unit. Project staff in the requisite professional disciplines such as engineering will be responsible for these checks and inspections. It has been stated that the monitoring unit must be seen as part of the management team; it is now clear that monitoring itself is part of the function of all members of the management team.

Similarly, when fertilizer, insecticide, or equipment is being made available, the delivery system must be monitored to ensure that it is reaching the intended recipients at the right time, in the right quantity, and at the right price. In a credit scheme provision should be made in every case for brief documentation of the criteria used for selection, the period taken for decision, and the numbers rejected, as well as the numbers granted.

Vigorous, continuous, input delivery monitoring of this kind will be effective only if the monitoring staff and managers spend a lot of their time in the field and receive the support of the field supervisory staff. They should keep daily logs of their activities, and submit regular returns showing the proportions of time spent in the field. Each level of management requires inspection by its superiors: those carrying out the monitoring functions must expect to be checked in the same way.

The question of timeliness can be treated under two heads. First, were the inputs made available at the time planned? This aspect can often be covered by monitoring the administrative records and staff reports of physical performance. It should be possible from these sources to show that not only were x tons of fertilizer delivered, but also that they were delivered to a specified destination at a certain date. Difficulties may arise when the goods are passing through the hands of more than one agency and there is—as is often the case—agency competition or rivalry. Any discrepancies in reporting must be closely investigated by the monitoring staff.

The second question is, if the inputs were delivered at the time planned, was this the appropriate time in the light of current conditions? The timing of planting for example, may have been advanced due to the early onset of the rains, whereas the delivery of seed may have been timed for a more normal rainfall pattern. In such a case, the monitoring system will be put to its greatest test. Was it able to detect the problem in time to allow management to remedy the situation?

Achievement of project objectives will involve a sequence of planned operations, each of which has to be completed at a specified time so that subsequent stages can proceed. A number of procedures for management, such as critical path analysis, have been developed in recent years. If one, or more, of these procedures are adopted by management the monitoring system must be organized to service such a procedure. If no such method is used initially, monitoring may indicate the need for their introduction if delays and breakdowns are beginning to occur.

2.6   The Role of a Monitoring Unit

As stated above, the monitoring function is a management responsibility and is one of the responsibilities of each member of the project staff. A monitoring unit assists in the operation of the monitoring system in the following ways:

(a) identification with management of the targets for project implementation and the indicators to measure progress against these targets;

(b) collation, summarization, and dissemination of the information flowing from the various agencies and staff engaged in implementing the project;

(c) analysis of the administrative files and records that pertain to the project implementation;

(d) collection and analysis of data from beneficiaries that are needed to supplement the available records and reports;

(e) maintenance, in a retrievable format, of the various data series over time as an aid to later evaluation; and

(f) preparation of reports that highlight the findings of the various analyses and, to the extent appropriate, present a range of logical options requiring decisions by management.

The type and frequency of reporting is discussed in Part 10. As discussed there, the communication of information will range from the frequent updating of a graph on the manager's office wall, to substantive 'state of the project' analytical reports. Flexibility must be incorporated in the organization of the unit's work so that it may respond quickly to major changes in the situation as they occur. A drought, a change in government policy, or other unexpected occurrence will require special analysis and emergency reports.

In order to report with an agreed frequency, deadlines must be set for the data or information to reach the monitoring office. The deadlines will depend on the dispersion of the project both geographically and over agencies, and the efficiency of the communication system. It is important to set deadlines that are sensible and which can be met. It is better to have a fifteen-day period for submission of routine returns (after the end of the reference period), which is adhered to, than a ten-day one which is not.

The successful integration of the unit within management requires that it does not appear to be acting as an 'overseer' or 'inspectorate' of project activities. Its role in making recommendations or participating actively in the decision-making process will depend on the type of management structure. But much will also depend on the personalities of those involved and their success in acting as an information service agency for the project.

2.7   Concluding Comments

This introductory Part has emphasized

(a) monitoring is different from evaluation;

(b) monitoring is an integral part of management;

(c) the monitoring unit must be a service unit for management, closely identified with it;

(d) prosaic, infrastructural records, both administrative and financial, are at the heart of a monitoring system;

(e) information from project staff is an important input for monitoring purposes;

(f) supplementary data collection must be used to fill information gaps, not to duplicate existing sources; and

(g) the key to success is a combination of timely action, concise reporting, and flexibility in response to unexpected developments.

PART 3:
EVALUATION: CONCEPTUAL ISSUES AND BASIC OBJECTIVES

3.1   Introduction

Experience shows that the management aspects of monitoring need the particular emphasis put on them in Part 2. This does not mean, however, that evaluation can be ignored within an individual project context. As stated in Part 1, evaluation aims to determine whether the project objectives have been, or are being met. Ideally, evaluation not only aims to quantify the achievements, but assesses the role of the project in bringing them about. Further, unexpected changes, both beneficial and detrimental, are looked for and their possible causal relationship to the project examined.

The accumulated experience of socioeconomic investigations demonstrates that only rarely are these ideal objectives fully realized. This part shows why this is so and then sets out less rigorous objectives for project specific evaluation with indications of how these can be achieved. The logical background for establishing both change and causality of change is provided in order to indicate the difficulties of applying it in the evaluation of agriculture and rural development projects and to demonstrate why a number of the scientific approaches are not likely to be effective. The more limited procedures then suggested can be assessed in the light of these arguments.

The most exact method of establishing causal chains is the experimental method developed by the natural sciences. An examination of a scientific experimental model reveals the features that are crucial to its effectiveness, but which cannot usually be reproduced in project evaluation studies. The human sciences have developed a set of quasi-experimental techniques to adapt the natural science model to the social world. An alternative, also successful, is termed the *modus operandi* method, which works much as a burglar's standard method of crime works to reveal his identity to the detective. This method relies on common sense expectations about behavior generated within a society's daily life. An example in agriculture is when a new seed generally operates in a certain way, the expectation is that it will continue to do so. Results will be attributed to it unless other factors are seen to be at work.

The human sciences—the social sciences (including law) and history—employ and adapt both methods. The experimental model is common in social psychology. The *modus operandi* method is used by historians to establish a sequence of actions followed by an individual or a group (who was where and who did what), consider their probable repercussions, and so attribute responsibility. Often circumstantial evidence is used.

A related method is to evaluate the likely impact of an action by asking: "What would have been the likely course of events if X had not occurred?" This device is known as a counterfactual. The hypothetical counterfactual sequence of events, against which actual events are compared, is generated by a sequence of plausible causal relationships.

Two general comments are made. First, questions about cause and effect have occupied the minds of many of the best thinkers, past and present. This is reflected in a vast and continuously expanding literature. In this Handbook a "common sense" view of causation provides the framework for discussion.

Secondly, the human sciences have not been very successful in establishing causal laws of the kind found in the natural sciences. Among the reasons commonly cited are:

(a) the great diversity and variability of human behavior;
(b) the adaptive and reactive nature of human behavior; and
(c) the inability to experiment, and to replicate experiments, in most social environments.

Thus, it is usually impossible to establish, with full rigor, the causal chains that it is the job of evaluators to find. They will normally have to be satisfied with showing that it is "plausible" that the project has had the impact expected, and that there have been no substantial adverse side effects to offset its benefits. The question then is: "What is to be regarded as 'plausible'?" Such a question, however, by its very nature, has no single, simple answer. The rules, techniques and suggestions put forward in this Handbook, taken as a whole, provide a framework for deciding what is plausible in the specific circumstances of individual projects.

Major development projects are not experiments, except insofar as they include research components, or else in the very general sense that relatively self-contained interventions in social systems may be described loosely as "social experiments". Projects are planned and financed because there already exists a body of knowledge and experience indicating that certain activities can be expected to produce certain results—in other words, that they have a *modus operandi*. Evaluation in this context is to establish whether these expectations have been realized in a particular case. But this is, perhaps, to jump ahead in the argument: consideration must be given first to the scientific model.

3.2   A Scientific Experimental Model

The basic approach divides the subjects under study into two groups. One of these, labelled the treatment group, will be subjected to the predetermined causal stimulus or intervention, generally referred to as the "treatment" (it will be the project in the context of this Handbook). The other, labelled the control group, will not receive the treatment; its progress providing the basis for comparison. The situation is represented by the following diagram:

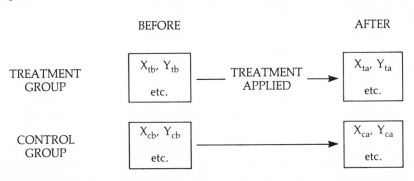

Effect variables (the features of state or behavior that the intervention is expected to change—denoted by X,Y, etc.,—in the diagram) are measured in the two groups, both before and after the treatment is applied. Changes in the levels of these variables in the treatment group are then compared with the corresponding changes in the control group: differences in these changes in the two groups are attributed to the treatment. If the treatment is of a kind that can be applied at different levels (for example, extension officers visiting once a month, or once every three months, or once a year), then there can be a number of treatment groups, one for each level. Such an extended system will usually give more powerful results, since relating changes in the levels of effects to changes in levels of treatment will clarify the causal relationship and can lead to more effective control in further applications.

When nonliving material is involved, a control group may not be necessary: it can be assumed that material not subject to the treatment will remain the same and no observations of it are required, but this constancy cannot be expected with human populations.

For inferences about cause and effect to be made with certainty, the two groups should have identical composition before the treatment is applied to one of them. Such identity is, of course, impossible, and so a looser relationship, "equivalence", may be attempted; for example, the study of identical twins. A restriction to this level of equivalence would clearly limit enquiries very severely; so a looser matching or "blocking" is attempted. Members of the two groups are matched according to their position on variables that are thought likely to influence reactions to the intervention. Matching variables used could be age, sex, socioeconomic status, level of education, area of holding cultivated, or size of household. It is a demanding task to obtain and maintain a successful match on one variable: to do so on several variables is very difficult.

One frequently used method of achieving equivalence is by randomization. This involves allocating subjects at random to the treatment and control groups, so that they are equivalent at an "average" level; with neither of the groups likely to react especially favorably or unfavorably to the treatment. Thus, in a fertilizer trial, land is divided into a number of plots, and the different fertilizer treatments will be allocated at random to them. (There are, of course, much more complicated experimental design arrangements than this simple random model, but they are not relevant to the argument here.)

The restricting nature of this model should be realized. It is not just a question of obtaining a random sample of subjects: those subjects must then be randomly assigned to the treatment and control groups. In several schemes in the past, where random allocation was feasible, it was not done because people were allocated on a basis of need rather than at random: for example, the poorer children have been put into the treatment group to get free milk, i.e., the treatment.

Random assignment will usually be impractical in agriculture and rural development projects. It may be that a project will include certain villages and exclude others nearby which are similar in some respects. But it is very unlikely that the villages in the project have been selected at random (although commencing work in a part of the complete and homogeneous project area may be considered a close approximation). The fact that *they* have been chosen usually means that they are different in some way from those excluded; and it is often expected that these very differences will influence reaction to the intervention made by the project. If the scheme is designed to affect the poorer farmers, then a control group of the better-off farmers is not effective. Nonrandom groups provide information of value (see Section 3.3) but conclusions drawn must pay regard to their nonequivalence.

With living subjects, there will be natural change over time, a maturing (or maturation effect as it is called), and this will complicate the assessment even if the groups are equivalent at the outset. Moreover, the environment will also influence the variables selected to measure project effects. Some of these environmentally induced changes may be larger than those anticipated from the project. In agriculture and rural development, fluctuations within the year and large variations from year to year are common, due primarily to climatic variability. The "message" or "signal" of the project will often be obscured by the welter of the effects of these other changes. Nor is it likely that the two groups will experience exactly the same sequence of environmental changes over time and space. There may be a new market in one area or a new road in another. It is then difficult to partition differences between the treatment and control groups into those which are due to the treatment and those which are due to exogenous causes.

In the experimental model the treatment and control groups can be effectively isolated from each other as well as from the outside world. Such isolation is not possible in most project circumstances. One aspect of this is imitation. If some farmers adopt new practices, others may copy them. This imitation is often part of the project design; even when it is not, it is undesirable on ethical grounds to prevent it. In all cases its existence is important in any final evaluation of the project impact.

Human reactions may upset the experimental conditions in other ways. In medical trials the control group is given a placebo—a pill or treatment expected to produce no physiological change—to allow for the possibility that results are being caused by the psychological effects of being given any treatment, rather than by the substantive effects of the particular innovation under trial. It would, of course, be ridiculous to consider a placebo in an agricultural context: visits by extension workers which did not transmit any useful information would be difficult to arrange, a waste of resources, and counterproductive.

A similar situation affecting the treatment group is called the Hawthorne effect, named after the factory in which experiments showed that productivity was stimulated by the circumstances deriving from the group being a focus of interest, rather than by actual changes in conditions of work. It is conceivable that a group of farmers, previously neglected, might respond favorably to almost any project involving them, no matter what its content.

Yet another confounding factor—common in medical trials and guarded against by the "double-blind" procedure—can arise. This occurs when the administrators of the treatment give more attention to those receiving treatment than to others. Thus, an extension worker may give more detailed and careful advice to farmers who are receiving credit than to those who are not. It would be undesirable to discourage keenness or enthusiasm in putting a project into effect, but the danger is that part of the recorded effect may be due to this general encouragement rather than to other specific project inputs. Evaluation must include consideration of what happens when intensive pressure, which is part of the project package, is withdrawn or reduced.

These effects could, of course, be considered as project benefits, arising from the project organization rather than its content. But evaluation will normally have as its objective a measure of the effectiveness of the specific inputs provided.

In scientific experiments there is usually little loss from either group. In human studies the respondents may not be prepared to continue their participation as the tedium of interviews builds up over time, or they may migrate. The composition of farmers involved in some PIDER schemes (in Mexico) has continuously changed, although the changes have been concealed in relatively stable total population figures. Losses of this kind are particularly damaging if they occur at different rates in the two groups. Since the final project impact cannot be assessed until long after its completion, it is extremely difficult to maintain effective groups, even if it is possible to set them up in the first place. In addition, the longer the implementation period, the greater the possibility of substantial maturation and environmental changes, affecting the groups in different ways.

Summing up, therefore, it is not usually possible in development projects to obtain equivalent groups or to achieve random assignment; the demands of matching are exacting; and, even when equivalence is obtained at the beginning of the project, differential dropouts will almost certainly introduce nonequivalence during the project. Methods other than the scientific experimental model are therefore needed. This leads to consideration of quasi-experiments, developed for circumstances when cause/effect relationships are looked for in situations where control and treatment groups are nonequivalent, or there is no control group.

3.3   Quasi-Experimentation: The Nonequivalent Group Design

Quasi-experiments have been defined as "experiments that have treatments, outcome measures, and experimental units, but do not use random assignment . . . comparisons depend on nonequivalent groups that differ from each other in many ways other than the presence of a treatment whose effects are being tested. The task . . . (then) . . . is basically one of separating the effects of a treatment from those due to the initial noncomparability between the average units in each treatment group."[1]

There are two basic designs for quasi-experiments. The first is a "nonequivalent group design". It retains the idea of a control and a treatment group (or a number of treatment groups), illustrated in the previous section, but without random assignment. Matching or blocking is done at the analysis stage after the groups have been set up in the light of the situation in the field. Although matching is postponed, the intention to do so later must be taken into account at the planning stage. The choice of the data to be collected will determine how far the available analytical techniques can be used in the later stages of evaluation. Thus variables that can be used later in an analysis of covariance, or for standardization, must be identified before starting data collection. The complicating factors discussed above that can weaken causal inferences even when randomization or detailed prematching is possible will, of course, continue to operate in nonequivalent group designs. It is difficult to judge in advance whether they may be more or less damaging.

The additional problems of nonequivalent group designs relate to establishing whether or not causes other than the treatment may have been responsible for differences in the changes of the two quasi-experimental groups. The factors of environmental change and maturation are likely to be aggravated with nonequivalent groups. Their nonequivalence may result in their being exposed to different developments in socioeconomic change. If there is a series of observations, rather than just "before" and "after" enquiries,

---

1. T.D. Cook, and D.T. Campbell, Quasi-Experimentation (Chicago: Rand McNally, 1979).

it may be possible to identify (and therefore possibly allow for), any disturbing environmental events. Differential maturing is particularly likely if the treatment group is self-selected, since it then probably consists of the more energetic or able, or of those who are in a position to benefit most from the treatment.

One version of this problem is the tendency for the rich to get richer, and the poor, poorer; or "nothing succeeds like success." In this situation, even if a treatment group—in this case, the poor—remain poor (or, indeed, are worse off), in relation to the richer control group, it may be argued that its position would have deteriorated even more *but for the treatment.* Such an argument may be true: but it is difficult to demonstrate. It is an example of a counterfactual argument, and needs a strong data base from which to establish with a reasonable degree of confidence what the trend would have been without the treatment. This is an example where a joint approach of also using time-series information is necessary.

When groups are equivalent, the same indicators and techniques of measurement will work similarly with both. So long as the nonequivalent groups are not too diverse this situation should continue to apply. However, if there are particular differences between the groups along some variables, additional problems will arise. Indicators measured by scales are especially hazardous, owing to "floor" or "ceiling" effects which arise because the extremes of the scale operate in a different way to the rest of the scale around the middle. For example, if one group is, at the outset, near the top of a scale (that cannot exceed 100%), its rate of growth using this indicator will be very different to a group starting from a much lower scaled value.

If the change likely to be effected by the treatment is not large in relation to changes being generated by the ordinary interplay of social and economic forces (and this is the case with many projects), then the size of the sample required to measure the impact will be large. (A more detailed discussion about sample sizes is given in Part 7.) The control group, may have to be of about equal size. Using a considerable share of available resources to collect data from those who are not of direct interest to the project will often meet with resistance. There is a natural tendency to concentrate attention on the treatment group. This may be offset if the control group will be beneficiaries of a later phase of the project; the "control" data will, at a later stage, become useful "baseline" data.

However, in quasi-experimental designs it may be possible to use an existing sample group as a control. There may be other sample enquiries going on in the neighborhood of the project. If the project has a wide area of application, data from a regional or national enquiry may be available for control purposes. If there are ongoing investigations which are only partially relevant, it may be possible to have them extended to provide additional useful information. Such a procedure will almost certainly be less expensive than the setting up, maintenance, and observation of a special control group, and will release resources for the intensive study of the treatment (project) group or associated case studies. An additional advantage of using an existing survey group in this way is that data for the group may extend back into the past. Further, questions about indicators and methods of collecting data will already have received attention, and this experience can be used.

Not every existing enquiry provides a useful control group, but the growth of survey capability in many developing countries, and the increasing number and extent of enquiries into rural households suggest that this possibility should be investigated.

When special treatment and control groups are set up, data will normally be recorded for each individual in each group, and may be analyzed at this level. But sample groups may not maintain constancy of membership over the project period; in this case it is particularly important for the analysis to include a detailed study of within group variation as well as between group differences. When a sequence of groups of changing composition is being used, within group variation is one of the few tools left for disentangling causal factors.

3.4 Quasi-Experimentation: The Interrupted Time-Series Design

The second type of quasi-experiment is an interrupted time-series design. In its simplest form this merely involves a before and after comparison of a treatment group. If the treatment is spread over a period, as in most agriculture and rural development projects, the before/after comparison will be strengthened if a mid-term picture is also obtained. As the name "interrupted time-series" implies, however, this design works best when the data are available in a regular time-series and the treatment is a sharp shock of relatively short duration at some point in the series; for example, a change in pricing policy for a crop, the change coming into effect at a specified date. This latter condition militates against the model in many projects where the inputs to change are distributed, of necessity, over a long period.

This simplest design is the most practical proposition for most agriculture and rural development projects. If adopted, every means of strengthening what is a weak design must be incorporated. How this may be done is discussed below.

Successful identification of causal chains will be more likely if the interrupted time-series design can be combined with the nonequivalent group design introduced above. If no control group can be identified, or only a rudimentary control from other sources of survey data is likely, the beginning of the time-series has to act as a control against which developments in the treatment group will be assessed. A pre-implementation survey is, in these circumstances, essential, since this is the only way of ensuring that the "before" situation measurement is not "contaminated" by any project effect. If the project involves large-scale construction, whose benefits will not come on stream until the engineering works are completed, then the base point for the time-series can be carried out during this phase of the project. However, the construction, and the employment it generates, will almost certainly have some substantial effects. For example, in one series of projects the benefits of the employment provided by well boring were so welcome to the farmers that they preferred to seek continuing employment in extensions of the construction program, rather than return to their holdings and use the newly available water for increasing production, as had been intended.

Any inferences about cause in this nonexperimental framework will depend on detailed information about the progress of the project obtained from the monitoring data. Such a combination provides the basis for the *modus operandi* approach mentioned at the beginning of this part. The description of the sequence of events occurring during the project will go some way to explaining the effects and towards arriving at a plausible conclusion concerning the cause.

The interrupted time-series is used to detect changes in levels and rates. These changes are likely to require time to diffuse through the treatment group or community. Even when the project provides a sudden intervention, e.g., the water starts flowing through the irrigation canals, the ultimate benefits may only accrue after some time. Thus, the period of postimplementation evaluation—the "after" survey—must be sufficiently long to measure the change and also the *persistence* of the change. The longer the time-series, the more possible it will be to make allowance in the analysis for seasonal and cyclical changes.

Special care is necessary to ensure that the survey methods and data collection quality are maintained throughout the period of reference. Changes in these may result in spurious signals emerging from the analysis.

3.5 Investigating Relationships Through Case Studies

The discussion to this point has assumed that the examination of change and causality of change is conducted through the medium of surveys with the groups selected in such a way that inferences can be made about the population at large. Detailed examination of "representatives" of experimental and control groups is possible if the number of cases is kept very small. The necessary equivalence can be achieved and the depth of study that a skilled observer can achieve at this microlevel enables the determinants of differential change to be identified, at least within the reasonable or plausible confidence limits suggested earlier.

More will be said on the circumstances under which it is appropriate to employ sample surveys or case studies in Parts 5-7. But it is emphasized here that case studies may be the evaluator's most appropriate medium for investigating causality. They are certainly simpler to organize than the longitudinal data series with its demands for an appropriately timed baseline and a continuation beyond the implementation period.

3.6   Practical Conclusions

Starting from the framework of a strict experimental model and taking account of the difficulty of applying it as used in the natural sciences, this Part has attempted to progress towards a set of practical objectives for project specific evaluation.

The following summarizes the guidelines offered:

(a) Adopt designs predicated on the use of matched control groups only when the exacting conditions for their selection can be met;

(b) Search for and utilize looser control data from other enquiries, particularly national or regional surveys;

(c) Commission detailed case studies to probe in-depth specific phenomena that are postulated to be affecting project performance, utilize the findings of these case studies to help draw "plausible" conclusions on relationships;

(d) Rely on time-series analysis only if a baseline is commissioned in due time, with provision for follow-up enquiries;

(e) Select indicators that can be measured accurately by available data collection methods;

(f) Maintain comparability of the data over time by consistency in survey methodology, quality of data collection, and analysis; and

(g) View the monitoring and evaluation system as a whole, using the interplay between the different reflections it offers on project progress to ensure that conclusions of routine statistical analysis are substantiated by observed evidence.

PART 4:
THE SELECTION OF APPROPRIATE VARIABLES AND INDICATORS

4.1   The Criteria for Selection

The opening parts of the Handbook have set out the basic precepts in the design of monitoring and evaluation systems for agriculture and rural development projects. In this Part, the crucial issue of the variables or indicators to be measured within these systems is addressed. Clearly, in view of the wide diversity of projects and the differing levels of resources available for data collection, it is impossible to set out a definitive list of variables to be measured: what is appropriate in one project may be inappropriate in another, and what is possible to measure in one environment may be impossible in another. On the other hand, the purpose of the Handbook will not be served by merely listing all those that have been suggested in various guidelines or used in a number of projects with greater or lesser success. A review of the accumulated experience to date provides a body of evidence that enables one variable to be recommended as likely to be useful in the majority of cases, or another to be flagged as one which has proved difficult to collect or to relate to project objectives. Such an attempt is made here. Variables recommended should at least be given first consideration. Those that are criticized may be appropriately included—but with the prior knowledge that they will demand particular skill and resource in their use.

A distinction is sometimes drawn between a variable and an indicator, reserving the latter for derived ratios of variables, or transformations of the value of a variable into an index or scaled format. But the distinction is a blurred one: a variable may be chosen as an indicator because, in some way, its value represents more than its direct meaning in its limited context, e.g., infant mortality is a variable that has been widely used to represent much more than *infant* deaths, rather, it has often served as an indicator of the relative position of the society in terms of health development. So, to avoid confusion, the word 'indicator' will be used here to also mean 'variable', unless the point being made requires a distinction to be drawn.

The general questions to be asked while selecting indicators include

(a) Can it be unambiguously defined in the conditions prevailing?
(b) Can it be accurately measured in the conditions prevailing, and, at an acceptable cost?
(c) When measured, does it indicate 'the state of a condition'; in a specific and precise manner?
(d) Is it an unbiased measure of the value of interest?
(e) When viewed as one of a set of indicators to be measured, does it contribute uniquely to explaining part of the variation in the situation it reflects?

Before proceeding to consider appropriate indicators for measuring project progress towards each objective separately, introductory comments on these questions serve to set the scene.

(a) Unambiguous Definition

Many economic and social indicators that have been precisely defined in relation to a cash-oriented, industrial society are often inapplicable to small semisubsistence farmers in developing countries. Examples are: unemployment or underemployment, and income.

In a wage-earning environment, unemployment and underemployment may be capable of clear definition, although there are difficulties here too. The definition of these terms in relation to small farmers has been a source of great difficulty and is still debated in international forums and in the social literature. What is a full day's work in the context of family labor maintaining a one hectare food crop farm? What meaning does 'actively seeking employment' have in a situation when there is no possibility of finding any?

The income of a semisubsistence farming household is capable of an almost infinite number of definitions. Is food produced for home consumption to be valued, and if so, at what price? Changes in the size and composition of the livestock holding raise income versus capital asset problems. What is to be done about exchange of food between neighbours? Is farm income gross or net of input cost? A recent study concludes: "The economists who determine the value of family labor (if it is to be charged as a cost of production), and of family consumption have far more control over reported levels of income than do the farmers themselves."[2]

2. *Information for Decisionmaking in Rural Development: Report for USAID* (Washington, D.C.: Development Alternatives Ltd., 1978).

(b) Accurate Measurement

Even when defined, many indicators are difficult to measure when dealing with farming communities which are illiterate and are not accustomed to dealing in standard units of measure. The demographic literature shows how difficult it is to obtain accurately even such fundamental data as age or mortality.

Yields can be measured accurately by crop-cutting methods (see Section 4.4) but such methods require time and skill and, as usually applied, may only produce aggregate community estimates rather than farm-by-farm estimates.

Measuring income, however defined, is extremely difficult, and not for small farmers alone. The report quoted above concludes, after a review of many attempts to measure income, that "If . . . the purpose of income measurement is to capture statistics (mean values) at one point in time, remeasure income at a second point in time, and attribute the differences to the intervention of the project, then the present state of the art of statistical surveys will not generate income measurements with sufficient accuracy to validate such conclusions."

As income is such a prime requirement for most potential users of evaluation data, this is an important finding and will be considered further below.

(c) Specificity and Precision

These will tend to decline as one proceeds along the sequence of inputs, outputs, effects, and impact. The delivery of inputs within a controlled project area may be specific—a change in the rate of delivery reflects a real change in the implementation of the project. The output achieved as a result will be less so as yields may move seasonally according to climatic variations by more than the underlying trend due to the use of the inputs. Income as a measure of the effect of the project may be influenced to an even greater degree by exogenous influences outside the control of the project.

In terms of monitoring implementation, precision may be a practical criterion in the choice of indicators; for evaluation of results, precision may be impossible to achieve. Fortunately, however, when evaluation is undertaken, there should be at least a reasonable series of values of the indicator over time; used as a time-series, the *trend* in the indicator may be detected reflecting the trend in the condition it is meant to represent, with the confounding effect of seasonal and random influences allowed for.

(d) Bias

Clearly, it is desirable that an indicator should, whenever it is measured, reflect the true value of the condition it is representing. However, in measuring change over time, a bias in the absolute level at each point in time may not matter so long as the size of the bias remains constant, either in absolute or relative terms, for different levels of the indicator.

Thus, for example, the volume of a crop purchased by a marketing board will usually be a biased estimate of the total crop marketed (unless a monopoly exists), but changes in the volume purchased by the board may be a good indicator of change in the total volume marketed so long as the board's share of the total market remains constant. Similarly, a farmer's estimate of production may be biased, but if the relative bias remains constant, a proportionate change in this estimate over time will reflect accurately a true proportionate change in production.

The problem is whether such an assumption of constancy in absolute or relative bias is justified. In years of large harvests, the price on the open market may drop and farmers may sell most of their crop to the marketing board at a guaranteed stable price. In lean years, the farmer may take advantage of high prices in his nearby market and the sales to the board may decline proportionately much more than the decline in total marketed volumes. Or, farmers may increasingly understate their production as the true figure rises—not wishing to indicate that their prosperity has increased.

(e)   Contribution to Explanation of Variation

In selecting indicators for a monitoring and evaluation system, the objective is to put together a set wherein each indicator, individually and separately, explains a part of the overall variation in the condition being studied; and, together, explain a substantial proportion of the total variation. In a well-documented and quantified system, the contribution of each variable can be identified by statistical techniques such as multi-variate analysis—identifying the contribution made by each in a step-by-step manner and deciding on its inclusion in the explanatory equation according to its unique contribution. Such possibilities do not normally come the way of the designer of monitoring and evaluation systems. Nevertheless, previous studies or even intuition may identify an indicator that is so closely correlated with another indicator that the inclusion of both contributes little more than the inclusion of either one alone. Examples of pairs that are sometimes, but not invariably, highly correlated are

income	- expenditure
area planted	- quantity of seed used
literacy	- years of formal education
cash crop production	- cash crop sales.

There will be instances when the extent to which the relationship breaks down is itself of interest; thus poor correspondence between quantity of seed and area planted may be revealing of bad sowing practices. In such a case both indicators are, of course, necessary. Another reason for including two indicators that are closely interrelated relates to the sources of error in their measurement. If an indicator can be measured only approximately, or with a degree of bias, it may seem worthwhile to measure another related indicator in the hope that the two together will strengthen the conclusions that may be drawn, even though both are subject to measurement errors. This may be so if the sources of the error are different for each indicator, but is likely to create false confidence if the sources of error are the same, as both may then be equally misleading for the same reasons. Thus, if income estimates are biased due to deliberate understatement by respondents, it may be useful to collect expenditure if the errors in these responses are due to memory errors, rather than concealment, even though the true values of each are highly correlated. If both reveal similar orders of magnitude, more trust may be placed in the results.

4.2   Scaled Response and Proxy Indicators

Of the criteria listed above, the one of most practical importance from the designer's viewpoint is the ability to measure the indicator accurately, at reasonable cost on the scale required. Inability to do this may require a search for a proxy indicator that can be measured more easily and which appears to meet, or partly meet, the other criteria.

Consider first the example of indicators of the area under maize. The questions asked of a farmer can include the following, in ascending order of difficulty:

(a) How many plots did the farmer plant?
(b) How much seed did the farmer use?
(c) What was the area of maize (e.g., to the nearest tenth of a hectare)?

Either (a) or (b) may serve as a rapid assessment aimed at producing an indicator of whether the incidence of maize planting is above or below previous seasons. (a) will be particularly useful if the plots are cadastrally mapped and of approximate equal (and known) size. For many purposes, however, a more precise area may be required; but if the farmers do not know the area in these terms, a massive quantum jump in survey scale is required in order to physically measure the land using one of the acceptable techniques.

At the time of harvest, a set of options might include these.

(a) How many plots were harvested?
(b) How many bags were harvested?
(c) What was the production to the nearest 10 kilograms?

(a) is of little practical use, even for rapid assessment. At best, it will indicate how many farmers achieved no harvest at all due to drought, pests or whatever. The choice is between (b) and (c). The farmer may be able to reply to (b)—in bags or other local unit—with the implied margin of error due to recall lapses, varying moisture content, conversion from cobs to volumes of grain, etc. The resulting distribution for the group of farmers studied will be crude, but certain approximate parameters may be calculated from it. (c) will probably require a survey that includes direct weighing of the harvest in the required state and with procedures for standardizing weights to a fixed moisture content; without these precautions, the level of accuracy implied by the recorded figure is spurious.

It is this major direct intervention effort, necessary to obtain accurate measurements, that is the bane of surveys of small farmers who keep no records and who are unlikely to be familiar with standard international units of measurement.

Even if resources for a direct measurement survey are deployed, the time involved in data collection and analysis will usually make it inappropriate for rapid monitoring for, say, judging the success of the planting effort, or assessing the likely marketed volumes that may require project intervention. For monitoring purposes, the cruder scaled distribution approach may be necessary even if the more elaborate survey involving objective measurements is also mounted.

A different example is provided by certain social indicators. Consider indicators of nutritional status as a proxy measure of the general quality of life. As will be discussed below, simple measures of age, height and weight may be used to calculate indicators, particularly for children, that reflect not only the nutritional status of the individual but may, in appropriate circumstances, be used as a proxy for general well-being. The plausibility of establishing such linkages is involved here (see Part 3). If it can be shown that the anthropometric indices of children have improved relative to an established standard, is it plausible to infer further that their general well-being has improved also? These are difficult questions, but the extent to which such indicators can be used not only to reflect a direct state of a narrowly defined situation, but also to act as a proxy for a wider situation merits consideration at the design stage of the data collection systems.

In summary, therefore, it is recommended that before complex measurements are embarked upon in large-scale surveys, the following questions be answered:

(a) Will an approximate scaled response provide an adequate distribution to meet most of the analytical objectives?
(b) Can a proxy indicator, obtained from simple measures, be used in place of the direct indicator, with reasonable confidence that a change in one reflects a change in the other?

4.3   Indicators for Monitoring

Although precise indicators are not identified, the groupings below provide the outline of a recommended set of indicators for project monitoring.

It has been stated in Part 1 that monitoring assesses whether project inputs are being delivered, are being used as intended, and are having an initial effect as planned.

The first group of monitoring indicators includes

(a) financial disbursement figures;
(b) progress of physical construction relative to a predetermined critical path; and
(c) staff and equipment usage rates.

The source for these should be the administrative records maintained within the project. There are no difficult conceptual or data collection issues here; what is required are good record maintenance and rapid collation, summarization, analysis, and communication of the results to project management.

A second group of monitoring indicators include

(a) technical parameters requiring instrument recording, e.g., water flows, evaporation, soil moisture;
(b) environmental parameters, e.g., meteorological data; and
(c) economic parameters e.g., crop and livestock prices.

These are usually well identified and defined and are seen as useful by project management. Regular price recording of relevant standard commodities according to agreed procedures should be a feature of most agriculture and rural development monitoring. Seasonal cycles and annual trends in local prices affect project implementation. The not infrequent neglect of prices as an indicator is an example of overlooking the simple in favor of the complex.

Moving on to the first project/recipient interaction, there is a need for indicators of input supply; examples are:

(a) credit supply;
(b) direct farm supplies;
(c) extension advice; and
(d) educational, health or social facilities.

Data for (a) and (b) in terms of input supply should also be available from project records. Indicators will include numbers of recipients and the resulting percentages of predetermined targets; the distribution of these numbers by various quantity size classes; and the parameters of this distribution (mean, standard deviation, quartiles, etc.).

At one level, indicators of extension services are simple to devise with administrative records as the source of data (see Section 2.4). For example, the number of recipients of visits or training; the frequency of visits or length of training received by the recipients; the number (and ratios) of first-time recipients and repeat recipients. Of greater difficulty, both in terms of definition and measurement, is the derivation of indicators of the quality of delivery of extension input. A proxy for this is the rate of adoption of the advice or message conveyed. This is considered below under adoption rates.

The delivery of educational and health facilities can be measured not only by the number of buildings for providing these facilities and their staffing (see first group above), but also by the number of persons attending these facilities for training, examination, treatment, etc. In these areas, the supply of the input becomes confounded with the usage or adoption rates; for further details on which, once again, see below. However, to the extent that attendance at the facility does not necessarily result in adoption of the treatment, e.g., family planning practices, the attendance or input supply data may be fruitfully used as a denominator in usage or adoption rate calculations, rather than the total population 'at risk'.

Up to this point, the monitoring system requires little in terms of surveys of beneficiaries, rather, it involves the maintenance and proper use of available records. The next set of indicators, however, move the focus to beneficiary observation involving the use of the received inputs, and, in the case of credit, repayment performance. They include

(a) input usage rates;
(b) adoption rates; and
(c) repayment rates.

The usage of physical inputs by the beneficiaries will require a survey to determine the practices adopted, e.g., seeding rates, method of planting crops to which the input was applied, and so on. It is desirable to keep the questions on usage simple, and not to confuse it with farm management or detailed study; simple questions can be put to a sample of sufficient size to provide valid estimates for the total beneficiary population.

Adoption rates are, perhaps, underexploited as monitoring tools. They are of use as a proxy for the success with which a message is communicated through training or extension visits, and are also relevant measures of acceptance of medical benefits given at clinics. In the simplest example, a member of the population may attend a clinic and depart with prescribed pills, but that is not to say that the pills will, in fact, be taken as instructed or that the course of treatment will be maintained after the initial experience is evaluated by the recipient. Follow-up contact with a sample of the recipients of the advice or prescription is then required. This is a task for the monitoring unit.

Adoption rates of varying degrees of complexity can be conceived. The simplest kind require data relating to current year only. Thus, if:

T is the target number which should be reached,
A is the actual number reached,
D is the number adopting the recommended practice,

the following rates can be calculated:

$$\text{Performance index} = \frac{100\,A}{T} \quad \ldots \ldots (1)$$

$$\text{Penetration index} = \frac{100\,D}{A} \quad \ldots \ldots (2)$$

$$\text{Achievement index} = \frac{100\,D}{T} \quad \ldots \ldots (3)$$

The 100's are introduced to put the indexes in percentage form; and it will be noted that $(3) = \frac{(2)(1)}{100}$.

Note that, in order to calculate this range of indicators, and if the definition of adoption is unambiguous, no measurements or long interviews are required; merely a set of observations relating to receipt and adoption of the recommendation provided.

But, there are problems in the definition of adoption. If the recommendation covers a group of combined operations—to plant at a certain time and at a certain spacing—is the recipient an adopter if he follows some, but not all, of the recommended practices? It is possible to consider the advice as a set of recommendations and collect figures for each component of the set. If the recommendation is to use X kilos of fertilizer per hectare, is the farmer an adopter if he uses X/2 kilos per hectare? The answer may depend upon the purpose of the index. Using X/2 kilos may represent acceptance of the message and thus indicate effective work by the extension worker. If this is what is to be measured, the farmer is classified as an adopter. On the other hand, if the effective improvement of production is being assessed, and X/2 kilos per hectare is insufficient to affect production, the farmer should not be regarded as an adopter.

If appropriate records over time are kept, it will be possible to identify repeat adopters and 'dropouts', and even relate these figures to whether the message has been repeated each year, or whether some other practice is being emphasized. The related rates can then be calculated and plotted on graphs.

If a monitoring system provided nothing more detailed than some of these indices, it would tell project management a great deal about the reaction of the intended beneficiaries.

The need for, and usefulness of, such measures is clearly illustrated in a recent case where detailed studies of farming practices had continued for some years with little feedback to management. Belatedly, a quick survey concerning adoption rates was carried out in a matter of weeks which revealed that 70 percent of the original sample of recipients were no longer participating in the program. Another example of overlooking the simple in the desire for the complex.

For the monitoring of the repayment rates of credit similar rates may be used. Detailed consideration of this problem is given in a recent paper.[3] One common indicator is the collection index for an accounting period. It is dependent on the period and is of the form:

$$100 \cdot \frac{\text{Volume of collections or repayments for the period}}{\text{Volume of loans or installments due in period}}$$

Another commonly used index is the percentage of the portfolio which is in arrears, obtained as follows:

$$100 \cdot \frac{\text{Total arrears at end of period}}{\text{Total loan portfolio at end of period}}$$

3. J.D. Von Pishke, "Quantification of Loan Repayment Performance," Course Note 42, Economic Development Institute of the World Bank, (Washington, D.C.: July, 1979).

This may be misleading, as a varying proportion of the loan portfolio may represent new lending on which repayments were not due. A slightly more complicated index is suggested. It is a repayment index, R_t, derived as follows:

$$R_t = 100 \left(1 - \frac{\Sigma A_t}{\Sigma A_{max}}\right)$$

where ΣA_t is the cumulative total of balances in arrears (net of any prepaid balances) over the life of the loan or portfolio study horizon t extending from period 1 to period n, and ΣA_{max} is the cumulative total of balances in arrears which would have accumulated if no loan repayments were made. An Index number of 0 indicates total default, while 100 shows repayment exactly on time, or an overall performance consistent with timely repayment behavior.

In the range of monitoring indicators, the measurement of outputs have yet to be considered, (as well as the assessment of beneficiary reaction.) Output measurement and its corollaries are considered, for convenience, in Part 4.4 below. Assessment of beneficiary reaction, other than adoption and usage rates introduced above, cannot be summarized into simple indicators. Studies involving in-depth interviews with selected beneficiaries may reveal considerable insights into the attitudes and reactions likely to affect the project, but the interviews and the interpretation of the results require probing and analytical skills that are more individual in nature. Such studies are particularly necessary when the simple indices indicate that something is wrong. Problem identification is an important monitoring objective (see Parts 2, 5, and 6).

4.4.   Indicators for Evaluation

The design of evaluation systems has been discussed in Parts 1 and 3. The emphasis on direct survey methods and the likely necessity of viewing the analysis as an examination of a time-series of the data have been noted. Within this context, and bearing in mind the criteria discussed earlier in this Part, what indicators can be recommended for inclusion in the evaluation system?

For introducing the presentation in sequence, the range of indicators may be classified into three groups:

(a) output indicators, including disposal of output;
(b) economic indicators; and
(c) quality of life indicators.

Output Indicators

In most agriculture and rural development projects, outputs are expressed in terms of production; whether crops, livestock, forest products fish, etc. Where manufactured outputs are involved, the indicators need individual identification and lie outside the scope of this Handbook.

In general, quantification of production may be through direct measurement or the producer's stated estimate. If the producer keeps records, the problem is, of course, solved, but for projects addressed by this Handbook, that is rarely the case.

If direct measurement is adopted, the following common categories of production can be identified:

(a) seasonal crops—areas and yields per unit area;
(b) tree crops (including forestry)—numbers of trees and yield per tree (or area and yield per unit area);
(c) animals—live numbers, and off-take numbers; and
(d) animal products (including fish)—volumes, weights.

The indicators, therefore, are well defined; numbers, areas, yields, volumes, and weights. It is the measurement of these indicators that presents the problem. Livestock numbers by type, age and sex are feasible only if the herds are available for inspection; volumes and weights of daily or weekly milk yields or fish catches are very demanding in terms of repeat visits to the chosen units; areas and yield measurements require careful timing and take considerable time and expertise to execute.

The rule with such direct measurement of indicators must be that the time and resource that are required can be justified only if high accuracy is both needed and achieved. Unfortunately, this level of achievement can be guaranteed only if the survey is on a small scale, commensurate with the, probably limited, availability of skilled and experienced personnel necessary to make the measurements.

Frequent visits to take direct measurements make it practicable to collect other data that also require closely grouped multiple visits, such as input costs; e.g., labor inputs to achieve the outputs.

The conclusion, therefore, with respect to detailed observations and measurements is in line with the recommendation in Part 5 and elsewhere, namely, that studies incorporating such measurements are best suited to in-depth case study surveys rather than large-scale sample surveys: the experience with farm management studies, for example, bears this out.

The estimation of crop production deserves special mention. A well-documented and popular method is to conduct a crop-cut of a small portion of a known area of the plot in order to estimate the average yield per hectare for the crop. The problems that arise even with the objective method are described in Section 8.2. The alternative to direct measurements of production is to obtain estimates from the producer. A decision to do this requires positive answers to all the following questions:

(a) Did the respondent ever know the answer?
(b) If so, can he recall it at the time of the interview?
(c) Is he able to report it in standard units?
(d) Will he be willing to report it correctly?

If the land has never been measured, it is useless to ask for the area. If the cows are milked by a herdsboy and the milk consumed by the family, the respondent may have little idea of milk yields. Virtually, no one has even an approximate idea of the weight of firewood gathered by the family for domestic use.

Even if the answer was once known, the respondent may not be able to recall it accurately. The memory error is likely to be related to the length of the recall period and the definition of the reference period. These issues are discussed in Part 8.

The respondent may report in terms of local units of measure; if so, conversion factors must be calculated to standardize the results in kilos or litres.

For this method to succeed, the respondent must be willing to recall the required figure and to report it accurately to the best of his ability. He must, therefore, be informed of the purpose of the enquiry and his good will established.

In many cases, concerning production, farmers can, if the interview is well-timed and the questions appropriately phrased, give estimates that—within certain rounded limits—reflect reasonably accurately the true value of the output. The use of farmers' estimates as an indicator should not be overlooked. There is a tendency to assume that such estimates will be so biased as to be valueless, but there is some evidence from the Philippines and elsewhere that farmers estimates are sufficiently accurate for monitoring purposes.

To record, for example, that farmers' estimates of yields are below those obtained from crop-cuts, and conclude that the farmers' estimates are wrong is not necessarily correct. A difference of 15-20 percent may be explained by different harvesting techniques and biases in the crop-cutting method.

The decision on these alternatives can only be made on the spot, perhaps following a pilot study. But a preliminary general recommendation is:

In micro, multipurpose case studies use objective measurements of outputs to the extent possible; in large-scale sample surveys, farmers' estimates may serve as a proxy for the objective record.

Indicators of disposal of output present similar alternative choices; post-facto reporting or measurement at the time of disposal.

Gross volumes passing through the market channels as estimated at certain points in these channels are a measure of total sales, but they often cannot be analyzed according to source.

Moreover, estimates of household consumption of home production and informal disposals by way of gifts or exchanges require multivisit household budget surveys. In the project evaluation context, such surveys can usually only be justified on a case-study basis from which general patterns of disposal may emerge.

If a cash crop, e.g., cotton, is marketed almost exclusively through a single official agency, the records of that agency or its local depots may be the source of indicators of quantities sold, even down to the farm level.

Economic Indicators

The postulated economic benefits of a project are usually expressed in income terms; so income is the obvious choice as the indicator to measure; and popular practice confirms this. However, as indicated earlier, in the case of small farmers, the accurate definition and measurement of income is difficult; in fact it has rarely been satisfactorily accomplished.

Data on income from a cash crop may be possible to obtain, particularly if it is marketed through a limited number of traders or agencies whose records provide a check on farmers' responses. Prices of crops or other agricultural produce can be collected regularly at selected markets or buying centers and together with estimates of production can be used to provide independent estimates of cash receipts. In aggregate terms for a zone or stratum, such estimates of total income may be satisfactory for evaluating the economic benefit of particular crops or products.

If total farm or household incomes need to be measured, the difficulties become extreme. There are the definitional problems outlined earlier and response problems. Income generation is erratic, linked to sporadic marketing; the income of each household member is difficult to obtain from a single respondent; and informal activities, such as beer brewing or small-scale trading that produce income are notoriously difficult to cover in an income survey. Moreover, income is a particularly sensitive area of enquiry to most respondents in any environment; objective responses cannot usually be easily elicited.

The measurement of expenditure does not present some of these definitional and response problems. The experience of many budget surveys is that expenditure is recorded more accurately and completely than income; to the extent that in many reports containing cross-tabulations of one variable against income classes, expenditure classes are used as a proxy for these income groupings. Expenditure can be fairly accurately recorded if the recall and reference periods are appropriate. For casual daily expenditures, the recall period may need to be no longer than two to three days. Items such as rent, services, school fees, etc. may need a reference period of one month; and irregular major expenditures on household or farm items may be reported on an annual basis. Problems of prestige expenditure in an attempt to impress have to be considered—this is taken up in Part 8.

There may not be an exact equivalence between expenditure and income in absolute terms, but in many population groups, changes in expenditure reflect, fairly accurately, changes in income. Can expenditure then be used as a proxy for income in evaluating change?

The reason why the answer to this is only a qualified 'yes', is that expenditure surveys may require multiple visits throughout an extended time period in order to accurately assess total expenditure over that period. The maintenance of such a survey on an almost continuous basis is not something to be undertaken lightly. Given that major expenditures can be recollected over a fairly extended reference period, the problem is the frequency of visits

42

required to estimate average daily minor expenditures. One possibility is to conduct a one-to-two week survey (with several visits within the period), repeated at certain selected times in the year (reflecting, say, pre-harvest, post-harvest and other major seasonal highlights), and to average the daily expenditures recorded in each of these phases of the survey.

The use of expenditure in this way needs to be carefully tested in each situation, but may well provide an easier method and a more accurate indicator of income change than an indifferent record of actual income.

When survey resources are very limited, total major expenditure on housing, household furniture and equipment, and simple transport and recreational items such as bicycles or radios, may serve as at least a rough proxy for income change. Even changes in the inventories of consumer durables in households—data which can be obtained by question and observation in a relatively simple fashion—can be illuminating.

Quality of Life Indicators

Indicators of output and benefits, though difficult to define and measure, refer to identifiable and tangible phenomena. The choice of indicators of quality of life is even more difficult because the very concept is vague.

Aspects commonly agreed to affect quality of life include, food consumption, health, education, shelter, access to essential amenities, and life expectancy. They are not, of course, independent of each other:

Food consumption in terms of individual or household food *intake* is an even more demanding survey topic than income, and has been as rarely successfully accomplished, except in microlevel detailed dietary studies. Domestic food production and food purchases through expenditure surveys may be the nearest proxies that can be measured.

Accurate assessment of health requires physical and clinical examination by qualified personnel, which is usually practicable only on a small scale. Nutritional status merits consideration as a proxy composite indicator for health and food consumption.

An accurate determination of nutritional status may be no easier than assessing the level of health; this is particularly true of adults. However, the measurement of age, height and weight of young children enables a set of three indicators to be calculated (weight for height, weight for age and height for age), which taken together and assessed against the now widely accepted standards, is a good yardstick of the general nutritional status of the child population. In effect, such anthropometric indicators are proxies for the nutritional status of the child population and these, in turn, are used as proxies for reflecting the general levels of health and food consumption of the population under study. The great advantages of anthropometric indicators lies in the ease of collection of the necessary data on large samples and their reasonable sensitivity to change in the underlying variables of interest. Clearly, they do not mirror the population's specific health problems, but as indicators in the set of quality of life measures can be strongly recommended for more general evaluation purposes.

One rider needs to be added. Accurate age determination of young children may be difficult; although evidence is mounting that in many countries this is becoming less of a problem, demographers still detect biases in age reporting in many instances. If age is recorded inaccurately, the ratios of height or weight for a given age will be biased. Where such a problem exists, weight for height may be the most useful indicator of the three suggested.

Education is easier to define and measure. Literacy and years of schooling are measures of individual attainment, and percentage enrollment of school-age children in primary education may be a suitable community-level indicator. Literary and educational level of adults, however, measure the past and are unlikely to change within a short period. Their use as project evaluation indicators is, therefore, limited. Age-specific enrollment rates for children is of much greater usefulness.

Shelter can be assessed by measures of quality of housing; e.g., number of persons per room, and type of construction. It is, however, slow to change and is somewhat culturally specific. Expenditure on maintenance, improvements, and contents may be more useful.

Access to amenities and type of amenities are useful social/health indicators, but again are slow to change, unless the project includes a component to directly affect their availability.

Life expectancy—an indicator whose accurate calculation requires very large samples (assuming that vital registration is not universal)—is more likely to be feasible at the national level, through population census and very large-scale surveys, than at the project-specific level. Moreover, in most developing countries, the calculation of life expectancy relies upon sets of 'model' life tables and the use of stable population models. As a project-specific indicator, it cannot be recommended.

Infant and child mortality may be feasible indicators if the full population in the project area is enumerated and a household register maintained. Otherwise, these too require such large samples as to be impractical.

Attempts have been made to assemble various combinations of quality of life indicators into one single composite indicator: one such suggestion is to combine life expectancy, infant mortality, and years of education. There are serious difficulties in interpreting the results of such a composite which implies equal weighting of indicators with no conceptual framework for doing so. Moreover, the lack of independence between the indicators confounds the position further; patterns of mortality exist covering the whole span of life, and these patterns underlie the models used to derive life expectancies in many developing countries.

If quality of life measures are required in a rural development project, the following is suggested as a reasonable list:

—Child nutritional status (anthropometric measurements);
—School enrollment by age;
—Expenditure on shelter improvements and contents ;
—Distance or time to potable water; and
—Use of clinic.

These are tentative recommendations and need to be reviewed to suit local circumstances.

4.5   Background or Classifying Information

In the sections above, the prime focus has been on identifying a number of practical indicators of direct interest for monitoring and evaluation. For the use of such indicators to be meaningful, it is necessary, in almost every case, to classify them against explanatory variables. These, also, have to be measured. Thus, in a rural community, the survey, whatever its main objective, will include a questionnaire on the general socioeconomic condition of the respondent. Likely topics include:

(a) geographical location and community characteristics;
(b) household composition and demographic characteristics;
(c) farm size and type; and
(d) main economic activities or occupation.

Indicators of these topics are not without their definitional and measurement difficulties. There are almost as many definitions of a household as there are household surveys; the definition of employment is one that has defied many students of the rural scene; and measuring age still causes problems in many communities.

Some of these issues are referred to elsewhere in this Handbook and most are dealt with in detail in manuals and handbooks issued by the UN Statistical Office and other specialised agencies and institutions. The point to be made here is that, although important, the collection of the background information is secondary to the collection of the monitoring and evaluation indicators; the survey must not be designed according to the dictates of the former at the expense of optimising the collection of the latter. There have been cases where the major part of the enumerator training period was devoted to the problems of defining and collecting data which were not of major importance in meeting the prime objectives of the survey.

Above all, the tendency to include everything that *may* be of interest in cross-classifying the main indicators must be avoided. It is this tendency that leads to monstrous book-length questionnaires, and results in poor quality of data collection for the important indicators. One result of such an unfortunate design is a report that is heavy on cross-classifications of the background variables against each other, but light on tables of the indicators that led to the survey. The extreme result is that processing delays become so acute that *no* report is published.

Of course, in many cases, certain of the explanatory variables may be of crucial importance in understanding what is happening to the beneficiary population. Family labour constraints, for example, may be the critical factor influencing the population against adopting recommended procedures. In such cases, the inclusion of such a variable may well come within the purview of the monitoring and evaluation systems.

PART 5:
SOURCES OF DATA FOR
MONITORING AND EVALUATION

5.1 Existing Data and Data Collection Agencies

The initial identification of data sources and assessment of data availability by those responsible for monitoring and evaluation of the project will be greatly influenced by the results obtained during the project appraisal stage. This appraisal depended on the collection, examination and testing of the whole range of available data. It will have identified the more reliable series; and indicated, by the use made of them, how to make the most of less reliable information, and of ad hoc material. Frequently it will have found deficiencies of coverage which were not possible to remedy during appraisal, and which the monitoring and evaluation will now have to fill or compensate for. If the appraisal's identification of these gaps led to an official data collection effort to cover them, the initial monitoring and evaluation effort may be geared to establishing contacts with the agency concerned and 'following through' on any further review of these data.

For the appraisal of a project, existing data supplemented by the necessary degree of estimation, extrapolation and even intuition, are used to assess the current position and project expected results. Project monitoring and evaluation commences by reviewing the existing data against a different set of objectives. To what extent can existing data series be used for deriving indicators relevant for project monitoring? Does the body of preproject data provide an adequate baseline for a time-series for eventual project evaluation? If so, will the existing machinery for updating these data be sufficient to generate the required time-series? Can the available data series be disaggregated to the project-specific area with sufficient accuracy? And are the concepts and definitions used in the past consistent with those applicable to project requirements?

To the extent that the answers to these questions are in the affirmative, the necessary arrangements must be made to maintain the relevant data series or to collaborate with the responsible agency in this task. New data series and survey activity must also be identified.

Contacts, both formal and informal, with key local and central officials involved in statistical development are required. Some of these contacts will be easy to establish through the project team. For example, in an agricultural project, links with relevant officials in the Department of Agriculture should be possible, both through daily contact with the staff of the Department and through the coordinating mechanisms built into the project management structure. Getting to know centrally based statistical agencies and the individuals in key positions will require a special effort.

Sources of data in any country are numerous and diverse. Much may be unpublished, since many enquiries are never fully analyzed, or written up. The material will be of widely varying quality and assessing the quality is as difficult a job as finding the data.

The following is a list of institutions which commonly exist and are likely sources of information. Actual titles and the relative importance of the bodies for data purposes will, of course, vary from country to country.

(a)   National Statistical Office

This may be responsible for censuses and certain national series, acting in other areas mainly as a coordinating body, leaving detailed data collection to statistical groups in other ministries. It may, however, be responsible for carrying out all major statistical activities; if so, this clearly simplifies the task of keeping abreast of new series and developments, and close contact must be established. The national accounts section will normally have a detailed overall view of data sources and their relative reliability.

(b)   Ministry of Finance, Economic Planning Ministries, and the Central Bank

These are sometimes responsible for the calculation of national accounts. They will certainly include economists who use the available data and who are therefore in a position to indicate the sources and provide some continuing assessment of data quality.

(c)   Production Ministries

These will often carry out data collection themselves. Licensing and marketing information may be centralized. For certain commodities, detailed information may be available in the Marketing Boards which may be independent or quasi-independent of the Ministry concerned.

(d) Service Ministries

If social and quality of life measures are relevant to the project, the data sources and experience of ministries such as those of Health and Education should be drawn upon. Project quality of life measures may need to be beneficiary or household specific, but school and clinical records of the area, together with data from specific social surveys may be useful in more general ways.

(e) Official Research Institutes

Although research data may not be directly applicable to project beneficiaries, they provide a check or target against which project-specific data can be assessed, and are useful preliminary data for rapid assessment enquiries (see Part 6).

(f) Other Institutions

These include universities and their attached research institutes, international institutions, trade associations, and private market research companies. In particular, universities and development institutes are major sources of in-depth, microlevel data and are, potentially, major collaborators with the project in certain specific studies.

(g) Output from Monitoring and Evaluation of Previous Projects

Experiences in collecting and using data in other projects, even those operating in similar environments in the same country, are frequently overlooked, due to poor dissemination of the findings. Such experiences are not only of potential use as a source of background data, but may help in identifying indicators that proved to be both possible to collect accurately and of relevance in monitoring and assessing project progress. Too commonly, the same problems are encountered, and the same errors committed by each monitoring and evaluation unit in turn.

5.2 Use of and Compatibility with Nonproject Data

It is appropriate at this point to reiterate the potential, of a national survey program as a 'primitive' control group (see Part 3).

Rural surveys, even though designed with different objectives, may reflect general production and socioeconomic levels. Health and education data, if they exist, provide a general picture of the prevailing situation. Project-specific data can often be thrown in sharper relief by using these other broader data files as a 'backdrop' for comparative purposes. The relevant agencies may even be willing to expand or modify their survey design and content to improve the potential use of the data for such comparison or control purposes.

In one area in particular, serious efforts at coordination are indicated; namely, the standardization of concepts and definitions. Many potential comparative analyses between data sets are impossible because the underlying concepts and definitions differ. It is common practice for such fundamental definitions as household and holding (farm) to differ from one survey to the next even within the same survey organization.

In the monitoring and evaluation context, careful attention should be paid at the initial planning stages to adopting, wherever possible, such standard concepts and definitions as may have been developed. Departure from such standards is often due not so much to different objectives, but ignorance that such standards exist. Comparative analyses of evaluation data with other non-project data sets will be easier if the following definitions, in particular, are common:

- Household composition;
- Migration;
- Holding (farm);
  - parcel
  - plot (field);
- Treatment of crop mixtures;
- Occupation;
- Employment (particularly informal and part-time family); and
- Units of measurement.

The collaboration between national agencies and project monitoring and evaluation units should not be seen as merely a one-way process. As stated in Part 1, the available monitoring and evaluation resources within a number of projects in a country, add up to a very significant part of the total statistical resource of the country. Although the design of monitoring and evaluation surveys must be geared to servicing specific project needs, these needs will often be similar to those of national planners.

Identification of these common interests by both parties, i.e., project and national authorities, will be mutually beneficial. Indeed, such an identification process often serves to both broaden and focus project management's perception of its own needs.

5.3   Sources of Project-Specific Data

The contribution made by nonproject data notwithstanding, the main sources of data for monitoring and evaluation will be project specific. These may be listed as:

(a) administrative records;
(b) rapid observation;
(c) case studies;
(d) sample surveys; and
(e) census results for the project area.

These will be considered in detail in other sections, particularly Parts 6 and 7. They are briefly described here to show the linkages between the different sources and the monitoring and evaluation systems.

(a)   Administrative Records

In one sense, administrative records, if they include certain information on *every* beneficiary of a project, may be regarded as a census; i.e., a complete listing of the population under reference. But in practical usage there is a distinction.

Much of the information required for measuring project progress will already be included in existing project files. Examples are financial accounts, construction and service records, credit applications and disbursements, marketed production figures, school enrollment, and clinic attendance records. No survey should be commissioned without first investigating whether the required data are available from such project records. One project evaluation unit undertook direct beneficiary interview surveys to obtain information, most of which was available, more accurately, in the individual beneficiary files maintained by the extension and credit agencies.

The use of administrative records requires a system for their rapid collation and summarization in order to turn voluminous files into succinct, decision-oriented information. Often, it is this system, rather than the data, that is lacking.

(b)   Rapid Observation

This comprises the technique used when making a quick tour of an area in order to obtain a general picture of the situation prevailing. It involves recording what is observed along the itinerary of the tour, what is obtained through discussion, both with persons encountered casually and with those specially selected (either purposively or by random procedures). It can be an exploratory visit, or one on which a special report is required (perhaps an emergency), or one of a regular series. It may provide general information about an area of interest, or be very concentrated, providing data for trouble shooting in a crisis. An agricultural officer, for example, will periodically tour his district and write regular situation reports (or informally accumulate his impressions, sometimes without written reports). The methods of observation which, of course, have been used since the earliest administrations, have recently been called 'quick and dirty' methods; but, although they are quick, they do not have to be dirty in the sense of providing unreliable information.

(c)   Case Studies

The term 'case study' may be used in different ways. It is frequently used to describe a report of one particular case, event or project which is presented as a valuable item for study, as a guide or example, as with a medical case history. In this Handbook the term is used to describe a long and detailed field study of relatively few households or persons, which examines in depth their lives and behavior, or some selected aspect of them. Common types of case studies include community (or village) studies, trace studies, and detailed activity studies (for example, farm management studies). They may be exploratory studies, research oriented, using experimental or quasi-experimental designs to provide evidence in support of large development projects. Obviously, in such cases, the monitoring and evaluation of the resulting project needs to include the original case studies in its 'data bank.' During a project, case studies are useful in finding out what has gone wrong if a project hits a major snag. More generally, case studies provide one of the ways in which the unfolding of causal mechanisms can be studied, as discussed in Part 3. In many instances, plausible causal inferences will be possible only if more general survey data are supplemented by a case study, examining in detail the project impact on a small number of beneficiaries.

The case study is usually carried out by a single professional investigator (or small team), perhaps with a few assistants. The number involved is small, but of high calibre. They must also be dedicated, for the investigation may involve a long period 'in the field.'

(d)   Sample Surveys

Samples (either purposive or probability) are very adaptable. They may be for descriptive or analytical purposes; or—more frequently—for a mixture of both. They can range from a quick study involving one topic in one area, requiring just a single visit for a short interview, to a project-wide multipurpose study involving multivisits spread over a long period. Since the number of respondents is greater than in case studies, it is easier to achieve representativeness of the population, and thus to generalize from sample to population. On the other hand, the survey is less suitable than a case study in examining intentions and attitudes and the way these work through to behavior and actions.

(e) Census Data

A national census is justified mainly on the grounds that the data to be collected from each individual is required at a very low level of aggregation; i.e. for very small administrative subareas of a country. It follows that if a census file exists it should be possible to obtain tabulations that are specific to the project area—or a very close approximation to it. This is a specific instance where collaborative arrangements with the main statistical agency in a country can result in access to such special tabulations. However, a census is infrequent (commonly decennial) due to the massive administrative undertaking involved, and the range of questions it can cover is limited. If available, census results, disaggregated to the project level, provide a starting point for almost every conceivable social and economic study using quantitative data. The census also provides the frame for many sample selection schemes.

Censuses of particular subsets of the population such as businesses or large farms may also be relevant in the project-specific context. If the most recent census was taken several years prior to the project, the continuing validity of the data needs to be reviewed. Moreover, assessment of the quality of census data is required. If a technical assessment is not already available, the monitoring and evaluation team will have to carry out its own if it has to make crucial use of census results.

5.4   Case Studies or Sample Surveys?

As already indicated, and as reiterated in Section 5.5 below, the appropriate answer to the question in this heading may well be 'both'; plus use of the other sources described above. Nevertheless, for particular data collection enquiries, the issue often facing the survey designer is whether to recommend a microlevel, in-depth study or a larger scale sample survey that allows more rigorous inferential possibilities. In extreme cases the choice may be clear. Detailed, multivisit observations involving interrelationships between the whole complex of farming practices are difficult (and expensive) to control on large samples; conversely, adoption rates measured on the basis of very few observations may not reflect overall adoption rates of the target population. But in many cases the choice is not a clear one. Obviously, availability resources is a major factor, but there are technical grounds on which the decision may also rest; the following classification may serve to illustrate these.[3a]

The criteria shown are the *scale* of the enquiry in terms of the phenomena to be investigated and the geographical coverage desired; the type of *interview* to be used; the method of *observation* and *measurement* required to collect the data of interest; and the *frequency* with which the collection has to be made.

3a. The approach described in this and some subsequent sections dealing with survey techniques is based on D. J. Casley and D. A. Lury, *Data Collection in Developing Countries* (Oxford: Clarendon Press, 1981).

For each of these, three classifications are shown, somewhat arbitrarily chosen, to represent points in what may be a more subtly changing distribution. For example, questionnaires may contain both open-ended and closed questions; a complex measurement in one environment may be simple in another; and so on. But the typology offered provides some clues in the choice between case studies and sample surveys.

CRITERION	CLASSIFICATION		
	A	B	C
1. Scale of enquiry	Phenomenon of interest rare and clustered.	Village or community level. Specific site or institution.	Phenomenon of interest widely distributed throughout area.
2. Interview type	Free-ranging; unstructured.	Open-ended questions; attitudinal studies.	Closed and/or structured questionnaire.
3. Observations and measurements	Technical, requiring professional skill.	Accurate and detailed.	Simple counts and measures.
4. Frequency	Continuous or very frequent.	Multi-visit over year.	Single visit.

Enquiries involving any classification including one or more (A) types should be based on case studies. Rare phenomena, free-ranging interviews, professional examinations, and continuous observation—none of these are well-suited to large sample surveys due to the demanding nature of the enquiry in terms of identification of appropriate respondents, and time and skill required of the investigators.

Enquiries involving classifications including (1B) together with (2B) are also best suited to case studies. Indeed, almost any combination involving (2B) would indicate a case study as the first option, for enquiries involving any appreciable number of open-ended and attitudinal questions are not easily controlled if large numbers of enumerators are involved—differential enumerator/respondent biases become potentially very serious.

The sample survey is well suited for classifications of type (1C) with (2C) together with combinations of (3B) or 3(C) and 4(B) or 4(C). If 3(B) is included the measurements need to be taken with great care and accuracy, and the field teams require especially careful training if a sample survey is to succeed. Multi-visit sample surveys (4B) are feasible but expensive to operate. A sample survey mey also be appropriate for classifications of type (1B) with (2C).

The classic sociological case study may be described as:

Type (1B) × (2A) × (4A)

A farm management study may often be:

$$\text{Type (2C)} \times \text{(3B)} \times \text{(4B)}$$

This combination may be suited for a sample survey (see above), although the need for accurate and detailed measurements together with multi-visits make it a difficult and expensive operation. But, frequently, farm management studies veer towards Type (2B) x (3B) x (4A). This combination is more suitable for a case study approach.

A study of a particular health problem may be:

$$\text{Type (1A)} \times \text{(2B)} \times \text{(3A)} \times \text{(4C)}$$

This demands a case study approach. But note that a nutrition survey involving anthropometric measurement of children may be:

$$\text{Type (1C)} \times \text{(2C)} \times \text{(3C)} \times \text{(4C)}$$

and lends itself to a sample survey approach.

Enquiries requiring the identification of particular types of respondents at a particular location, e.g., market trader interviews, are of Type (1B). If Type (2A), i.e., free ranging, unstructured interviews are involved (see Part 6), the enquiry is best handled on a case study basis. However, a regular but simple price collection is of:

$$\text{Type (1B)} \times \text{(2C)} \times \text{(3C)} \times \text{(4B)}$$

and this can be operated on a sample survey basis.

5.5 Combining Methods of Enquiry

The characteristics of the main data sources and the strengths and weaknesses of each have been described so as to indicate the issues to which they are separately relevant. In this context it is natural to 'set' one possible choice against another as in the section above. But this Part would be misleading if it did not finally stress that the main task in designing monitoring and evaluation systems is to *combine* the different methods in a total program that will use resources most effectively.

For monitoring, this combination will often be a mix of:

(a) analysis of administrative records;
(b) rapid assessment;
(c) sample surveys of adoption rates and
(d) problem-oriented case studies.

For evaluation, which has the monitoring data to underpin its efforts, the mix will frequently be of:

(a) causality-oriented case studies; and
(b) sample surveys over time.

The case study based on a professionally conducted intensive investigation will provide the kind of insight that cannot possibly be replicated on a sample survey scale of operations. These insights will identify key relationships and constraints and will generate hypotheses, thus focusing the range of enquiry more narrowly and suggesting indicators that *can* be collected by a sample survey. This combination of the results of case studies and of sample surveys will provide powerful evidence for assessing the plausibility of statements about the project impact and the manner in which it operated.

For example, a case study may show that only farmers with family labor equivalent of at least a wife and two children can cope with the essential tasks during some part of a proposed new cropping system. A sample survey can then establish the proportion of farmers with the required labor resources. It can thus be plausibly argued that the new cropping pattern is viable in areas where "x" percent of farmers display such characteristics.

Similarly, a case study may reveal the failure of a parastatal marketing organization to offer the guaranteed price for a food crop to small farmers, who are, therefore, dependent on the price offered by local traders. If this price falls below "y" units per bag, the farmer will be disinclined to produce a surplus for sale. A sample survey can monitor the price farmers are receiving. From these two enquiries the likelihood of farmers increasing their production can be plausibly derived.

These examples also illustrate staffing issues. A change of statistician during the sample survey should not make much difference (given general competence). The case study is, however, much more dependent on the experience of the chief investigator, and an attempt to substitute here may well result in a need to make a fresh start.

PART 6:
DIRECT OBSERVATIONAL METHODS

6.1   Direct Observation

The collection of data through rapid assessment (or reconnaisance) methods, and by case studies are jointly described as "Direct Observational Methods," since the responsibility for the whole work of designing, observing, analyzing, and presenting the results is concentrated in one person or in a small team. This is one of the key features that distinguish these methods from the sample survey; another, already emphasized earlier, is that the logical basis for formal probability inference is lacking.

This concentration of the whole process of data collection and analysis in one person, or in a small team working closely together throughout the enquiry, is a source of great strength, as has been described in Part 5. The bringing together of diverse strands of experience and information and the balancing of different kinds of evidence within a common intellectual focus, can lead to insights, the generation of hypotheses, and the formulation of problems with a range of potential solutions in an economical and effective fashion. Its success is, of course, heavily dependent on the qualities of the investigators, and the way these qualities match the subject of the enquiry.

The economic, social and cultural features of rural communities vary widely across the world; may indeed be very different within the boundaries of one nation. It is, therefore, dangerous to state that they have certain common features. There are, however, two characteristics of rural communities, reported so frequently and widely in rural studies, that they form a common background to be taken into account when planning direct observational studies.

First, there is usually a hierarchical structure in a village. The hierarchies will be based on different characteristics in different places, but will generally give positions of local power to the better-off farmers. Therefore, the views of those at the top will be more accessible and evident. Their contributions to, and domination of, discussions may produce the impression that there are generally held views on topics, problems and choices. Such unanimity may, however, not exist in fact. The differences will be difficult to uncover, since the poorer people, the landless, the women, or others may be unwilling to reveal their views readily in a group discussion in the presence of members of the 'elite.' Even in private individual discussions it might not be in their own interest to disclose opposition to village leaders, or indicate a different order of priorities.

Second, many villagers have an insider/outsider scale, ranging from close identification with their near kin living in the same community, through their kin in other settlements, neighbors in general, and then on to others who are 'outside' (ranging from local extension workers through to more senior officials who are less frequently seen).

Villagers will be naturally cautious in talking frankly to those they class as outsiders. Sometimes they say what they think the outsiders want to hear; or they may reproduce what they regard as the 'official' view, in the hope of receiving favors or special treatment from the enquirers or from their official connections. There may be intra-community or intercommunity conflicts which lead respondents to portray a view to gain support for, or approval of, their own side.

This is not to say that reliable information cannot be collected by outsiders on even their first visits to farmers or a village. There are, however, three important implications of this situation that affect the organization of monitoring and evaluation.

The first is the substantial improvements in quality of data and insights into situations that result from *continuity of good staff*. As observers make repeated visits they begin to lose 'outsider' status, they become aware of the villagers' deeper attitudes, intentions, aspirations, and priorities. They begin to grasp those implicit features of rural community life which are often crucial to the success or failure of the project.

The second implication is the need to adjust the time scale of the collection process to the kind of information required. The physical framework of an irrigation scheme may be readily observed. The ramification of the organizational system by which the water reaches a farmer's field in particular amounts at particular times will take longer to identify; and probing for causes of failure to provide community labor to clear silted channels will be a still more complicated task. A study of a project in rural Afghanistan notes that "the complexities of the water distribution system, water rights, negotiations between users at the head and foot of irrigation canals, could not be captured by survey technique (sic) . . . information on the potential for change . . . was generated through informal interviews, combined with in-depth discussions with knowledgeable local officials." In particular, case study methods adapting anthropological techniques will be needed to seek causal relations between input and effects; and to look for constraints affecting farmer behavior. Such enquiries need not take the very long time period common in classic anthropological studies. There is growing evidence to show that valuable 'applied' results can be obtained in shorter periods. Nevertheless, any agricultural enquiry may take more than a year to study those aspects (and they will be numerous), which require close observation of the sequence of operations over a complete farming cycle.

The third implication is that a conscious effort must be made to establish the points of view of the less vocal groups in rural communities. Thus women may be particularly concerned about water supply and fuel sources, and be in a much better position to judge needs in these areas, and present alternatives to male dominated strategies.

In Part 8, the way memory problems of respondents determine suitable recall periods in survey design is discussed. There is another aspect to the memory problem; namely, the manner in which the recall ability of the investigators determines procedures when using direct observational techniques. The rule is that information must be recorded at the earliest possible moment. Some information can be recorded as it is collected, for example, the result of an investigator's observation of physical aspects—the conditions of the roads, or of the crops in fields alongside the road. Sometimes it may be possible to record a discussion as it is taking place, similar to the way enumerators operate in a normal sample survey). However, obvious record taking is often undesirable in direct observational situations; it interferes with the development of informal relationships between investigator and respondent, which is one of the chief assets of this method.

The results of discussions must be recorded as soon after as possible, at least on the same day. There are three main reasons for this. The first is the obvious one of simple memory failure; compounded by the more than usual mass of information presented to the enquiring mind in the free-ranging observational situation. The second is the selective and structuring role of memory; items that do not fit in with an evolving mental perception of a situation are more likely to be forgotten, ignored, or distorted. If they are recorded at the time, they cannot be so conveniently disposed of. The third reason is the need to get into regular habits; if unrecorded observations accumulate excessively, recording becomes so burdensome that it does not get done.

6.2   Rapid Assessment: Observation and Enquiry

The role of rapid assessment as a source of data for monitoring a project was introduced in Part 5. Even though this is the most informal of the approaches to data collection, a certain amount of planning will be well rewarded in ensuring that the data, although obtained quickly, are of reasonable accuracy.

An observer's efficiency is determined by natural characteristics (sharp eyes, a good ear), experience, and prior knowledge of the relevant topics. Natural capabilities can be improved by practice in recording the results of observation. Classroom-type exercises may improve performance to some extent, but experience in the field is the best way (particularly if it is under guidance) and can only be obtained by 'doing' the job, not just reading about it. The rest of this section is, therefore, devoted to a discussion of the required knowledge of the topics concerned.

Investigators must start out with a clear idea of the topics they are supposed to be covering. A check list must be prepared beforehand of the physical features they want to identify and the questions they wish to discuss. Any such list should be regarded as open-ended, because not every eventuality can be foreseen. Preparing such a list, however, concentrates the mind; a study of the first draft may identify other related matters which did not come to mind at first.

The ground to be covered can often be best organized by using a simple framework of the following eight questions:

- Who?
- Where?
- When?
- What?
- How?
- For how long?
- How much?
- Why?

The questions may be simple: for example, "When was the last delivery of fertilizer to the distribution point?"; "What kind was it?"; "How much of it?"; They may be more wide ranging: "What factors prevent farmers from planting early, as recommended?"

Clearly, knowledge of the local language is an advantage. At least some knowledge of agriculture and livestock (an ability to recognize individual crops and a knowledge of cropping practices and the crop calendar) is required. Investigators lacking this, and making their first trips, should visit a nearby experimental station, or obtain help from the local representatives of the relevant departments before going out. Display of total ignorance will not encourage farmers to take enquiries seriously, and will not lead to any useful records of the state of crops in the ground. Similarly, it should be possible to obtain beforehand order of magnitude figures for the production of major items (e.g., milk yield per cow).

Knowledge of the average rainfall pattern (a main determining factor of the crop calendar), and of recent levels of rainfall is required. Recent maps of the area should be studied and taken on the trip; or a self-drawn map prepared. Any previous delimitation of statistical strata or of ecological zones should be entered on this map. Information about local units of measure must be obtained, and a standard basket or locally used container carried. Information about local marketing channels is required, and the places and times of the main regular markets identified and marked on the map.

The timing of visits needs consideration. The time of day will affect who will be found in a village and who on the farm: the day of the week will determine whether marketing operations can be observed; and the month will determine the farming activities in progress. Often, respondents can only be approached effectively very early or late in the day. Traders often arrive at markets very early, around sunrise. Farmers may only talk in a relaxed fashion late in the day, when their labors are completed.

Follow-up visits require less preparation, but background information must be corrected and amplified, and the checklist revised in the light of experience and expanded as new topics come under review.

Everyone develops an individual system of effective record keeping, but some rules are basic to the exercise.

(a) Record times and places of observation and discussion, and certain identifying features, for example, status or occupation of informants. If possible, maintain observations made on a journey on a separate travel log. Times, distances, and names of places can be recorded sequentially with the appropriate record of observations and events alongside. If standard numerical information at a number of sites is to be collected (for example, prices of selected goods at all markets visited), a previously prepared pro-forma will save time and provide a first structuring of results.

(b) Journal entries should follow a standard pattern designed to cover the checklist topics with a final, open-ended section. A regular pattern eases the burden of the work, and helps to ensure that nothing is overlooked.

(c) Develop a system of abbreviations for frequently used words. Records of actual words spoken are indicated by inverted commas ' . . . '. Separate out the observer's own comments and queries requiring further investigations—for example, by enclosing them in square brackets [. . .].

(d) Photograph key features, as long as this does not give offense. Annotate prints with dates, places and circumstances. A series of photographs over time are particularly useful. If promises to supply copy prints to respondents are made, ensure the promises are kept.

(e) A pocket calculator is generally useful.

(f) It *may* be possible to record discussions; but this may not be desirable since—as with writing notes during a discussion—free expression of opinion may be hindered. A tape recorder can, however, be used for keeping the daily journal entries when under pressure.

NOTE: Cameras, calculators and tape recorders must not be so much in evidence that they distract respondents and create the impression of the observer as a 'tourist'.

Field trips depend a great deal on logistic support. It is, however, notoriously difficult to obtain. Very few trips actually start at the time planned; a delay of a few hours at the start can result in a seriously curtailed program, or result in several extra days in the field.

Interview procedures are discussed in Part 8. One aspect is emphasized here; namely, the need to adopt an informal approach, free from traces of arrogance or of 'knowing best'. The presumption should be that if the farmers are following certain practices, there are good grounds for them. Local failure to follow project recommendations may be due to constraints which have not been appreciated by the project designers, such as lack of requisite resources, or failure to deliver support of the quality and quantity promised, at the time agreed. Thus, individual farmers may well not plant early if the date is in the period when village livestock are allowed to roam freely: if fertilizer was delivered late one year, there will be doubt that it will turn up at the right time in the succeeding year.

Poor farmers also attach a high notional cost to new arrangements, since the risk element always appears to be, and often actually is, high. There *may* be elements in current practice which are no longer required or justified, but 'traditional' is not synonymous with 'outmoded', and although a procedure is described in this way there may well be good underlying reasons for its continuance. Indeed, recently adopted practices, if successful and widespread, are often quickly labelled traditional: for example, contour cultivation in a hilly area in Africa was freely described as 'traditional' within a few years of its introduction. Food preferences are deeply embedded and are unlikely to change rapidly, particularly if pressure for change appears to be imposed from outside.

Experience shows that valuable information can be obtained if the investigator shows an evident willingness to listen to the farmer's opinion and tries to understand his position (see Section 8.4).

Locally resident government officers can be of great assistance, particularly those in agriculture and community development. It is always desirable, and usually neceasary, to give them advance information of trips to be made in their area. One reservation about information obtained from them should be mentioned. These departments will usually have a departmental 'point of view' to which their officials may subscribe, or which they think they have to be seen to support. They may tailor the expression of their views accordingly. Departments will, of course, also have specific activities under way, such as encouragement of new techniques; local officers, whose career success depends upon the adoption of these techniques (perhaps specified in target terms), are not necessarily unbiased reporters of what is actually happening. This issue is, of course, merely one aspect of a general difficulty that confronts all monitoring and evaluation work.

The main purpose of rapid assessment is to be *rapid*. This does not apply only to the method of observation. The whole process should be rapid, from the initiation of the trip to the final submission of a report and policy recommendations to the decision-makers.

6.3   The Case Study

As set out in Part 5, the term 'case study' in this Handbook refers to the detailed study of a small number of units, selected as representative of the group or groups relevant to the issue under consideration, but not necessarily representative of the population as a whole. The method is indicated when it is necessary to probe deeply into the interrelationships between people and institutions; to establish and explain current attitudes and beliefs, and to show why certain behavior occurs. Case studies are particularly appropriate when a *high* analytical content is required. They are free of the questionnaire and interview constraints of large sample surveys, nor are they subject to the shorter time horizon of rapid assessment, which prevents full rapport between investigators and respondents.

The period required for case studies will normally be long enough to require residence in or near the community under study. The investigators, therefore, have the opportunity of becoming accepted within the community. Sensitive issues can be discussed in a relaxed, informal manner and behavior observed. Investigators must, of course, minimize the effect their presence has upon what they observe, so that as far as possible what is there 'naturally' is recognized, recorded, and analyzed.

Thus the review of one evaluation effort states, "CIDER ought to explore ways of using its evaluation researchers more intensively by deploying its staff in the field on a more or less permanent basis . . . Maximum effectiveness can be achieved only if the current ratio of time spent . . . will be significantly changed in favor of more field time."[4]

Examples of project circumstances which indicate that a case study may usefully be commissioned include

(a) serious dislocations in the project which are not due to obvious physical upsets in construction, delivery or other logistic functions;

(b) wide variations in responses or achievements in different areas or sectors; and

(c) general, widespread underperformance or other evidence suggesting possible failure to structure project actions to beneficiary needs.

One common feature is the need to postulate and explore causal relationships. Regular project monitoring may identify the problem and will often provide the background for generating relevant hypotheses, but these need to be tested with more in-depth, accurately observed data sets.

Moreover, as already discussed in Parts 3 and 5, case studies will be relied on in many projects to supply the underpinning rational structure for judging the *plausibility* of competing causal arguments even if rigorous testing of their validity is not possible. Properly conducted case studies can bring into focus a whole range of project and related nonproject activities within a single behavioral framework and take account of system interactions.

The PIDER review referred to earlier states, "Such a study (which is multisectoral, but restricted to a rather limited area) should attempt to capture

---

4. Michael Cernea, "Measuring Project Impact: Monitoring and Evaluation in the PIDER Rural Development Project," World Bank Staff Working Paper No. 332 (Washington, D.C.: World Bank, June 1979).

the synergistic effect of several project components on the life, economy, structure, institutions and organization of individual communities."[5]

It can proceed in a flexible—often speculative—fashion, and still produce final results within an acceptable time span. As emphasized before, it will be most effective when it is supplemented by other techniques.

As the success of a case study depends almost entirely on the main investigator(s), the person(s) selected must be allowed considerable freedom to carry out the enquiry in the way they think best. Appropriate persons may be found from the project staff in the central evaluation unit (if there is one), from local academic or research institutions, or from consultants. It may be necessary to appoint two joint investigators since often the resources of both natural and social sciences (for example, an agronomist and an agricultural economist) have to be drawn upon. In some cultures it may be necessary to have investigators of both sexes. " . . . the need for an interdisciplinary approach is better recognized; an approach in which the sociologist/anthropologist works jointly with the economist and the agricultural researcher . . . appears . . . the most effective approach to understanding and affecting the . . . farming system . . . "[6]

The production of timely reports must be stressed from the beginning. Frequently, those most experienced and able in this type of study, once embarked upon it, find themselves drawn into unexpected avenues of enquiry, and pursue further refinements at the expense of meeting deadlines. Extending the period of fieldwork delays analysis directly (by postponing the effort required to be devoted to it), and indirectly (by inducing more complicated procedures). The final report may be a better document in the academic or research sense, but it is of little practical value since the time for making use of the study's findings is long past. Although the investigator must be allowed the freedom already mentioned, the progress of any case study in relation to other deadlines must be closely monitored.

In the past, one of the features of the case study type of enquiry has been its open-ended timing: admittedly, some interesting results would not have been obtained otherwise. It will, therefore, be a matter of fine judgement how far—if at all—any requested extension of the study period can be allowed. On balance, extensions should only be made in very special cases: it is too easy to suggest potentially interesting further relationships to pursue; and investigators may, as a result of their training, have an unconscious bias towards wishing to relate 'everything to everything!'

When commissioning a case study, a brief is needed for the investigator. Such a brief must specify the *topic*, *timing*, *scale*, and *subjects* of the enquiry. Required precision in some spheres must be combined with flexibility in others, since one of the reasons for using the case study technique is its exploratory function.

---

5. Ibid.
6. Ibid.

The specification of the *topic* will be relatively straightforward when the requirement is primarily descriptive, e.g. the provision of information about particular market channels, or an investigation into farmers' attitudes to some proposed or newly introduced facility. When the project is experiencing dislocations or wide response variations, the brief (even if focusing on the particular dislocation or variation) will need to be more open-ended; more of an outline of problems than a statement of specific information requirements. Project monitoring may have provided provisional hypotheses or explanations which will provide the starting point. In the final stages of evaluation, especially the detection of unintended effects, the brief must indicate particular areas of concern, and the general type of results expected.

In an irrigation project, for example, yields, which rose with the initial use of the water, levelled off at unexpectedly low figures. To explain the reasons for this failure to fully exploit the resources made available, a case study was initiated specifying as the topic the constraints experienced by the farmers as a result of the new circumstances brought about by the project.

The *period* allowed for the study must, as already stressed, conform to project needs. Generally, the more complex the issues, the more time will be necessary to investigate them. A search for interactions in the system, often in great detail, may be required. For example, planting dates for maize may be limited by planting dates for beans which may, in turn, be determined by the incidence of disease. There may be unappreciated interrelations between crop varieties and livestock. For example, in a maize project, the differential in varietal yields, when leaves were stripped from growing maize to feed animals, became a key factor in the assessment of returns to farmers. The whole system has, therefore, to pass under review. In instances of this kind, the actual *timing* of the enquiry as well as the period needs careful arrangement within the project framework.

Features of behavior representing risk reduction strategies are difficult to elucidate. Further, project failures may be due to inefficiency or illegality. It takes time to understand farmers' decisions and to uncover concealment or subterfuge.

The case study will need to delve into past events, and also observe behavior at the time it occurs. Labor inputs into the various stages of crop cultivation provides a good example: the frequency, extent, and time required for such tasks as land preparation, crop planting, weeding, treatment with fertilizer and pesticides, harvesting and post-harvest activities may need to be observed (or at least recalled over a very short time-span) throughout the agricultural season. This ability to observe behavior over time—one of the strengths of the case study—must be used with discretion. Such extended observation by skilled observers is an expensive operation and must be justified by the primary users', demand for detailed analysis. Too often the impetus for such studies comes from researchers who are not primarily concerned with the problems of the specific project. Evaluation case studies are not financed for research purposes.

The *scale or geographical dispersion* of the case study will also affect the length of time required (unless parallel independent enquiries are being mounted simultaneously). Scale may not be an issue in projects with a limited impact area, but in more extensive projects it is necessary to identify those factors which categorize project beneficiaries into suitable separate study groups. The case study units are then chosen from the groups. This selection should not be confused with stratified or clustered sampling (see Part 7). The aim is not to produce valid estimates for the population at large but rather to ensure that the case study includes examples that exhibit the main phenomena of interest. It is atypical or nonmodal groups that are likely to require particular attention. Some groups may not need to be studied. Randomness may not enter the selection process at all.

One essential is to identify and group in a meaningful fashion the *subjects*, so that what they do and how they arrive at the decision to do it can be studied and compared. The subjects for most agriculture and rural development purposes will be the farmers. Two qualifications should be made, however. First, although the farmer may make decisions at one level, other members of the family may decide operations at a lower level: for example, many cultivation practices are determined by women working on the family plot. The fundamental unit may be the household; "the family household is at one and the same time a unit of production and a unit of consumption . . . The sociology of this type of unit is intricately interwoven into its economy."[7] Secondly, when the household is split temporarily by migration of some of its members, the unit of decision making begins to break down. The investigation then becomes, perforce, more complicated. Where these aspects are important, the recommendations made here about methods for grouping farmers will need to be modified appropriately.

Factors affecting farmers' decisions may be grouped as follows:

(a) *General*

    (i)   climatic levels/variability
    (ii) pest/disease susceptibility
    (iii) community social/cultural
    (iv) marketing/credit facilities
    (v) government policy

(b) *Specific*

    (i)   resources: family and labor/land livestock capital
    (ii) individual attitudinal goals and response to change.

Clearly, classification by the general characteristics is the more straightforward. Much will be achieved by geographical divisions, although even an area with a homogeneous climatic regime may still need substantial sub-division.

---

7. Ibid.

For example, closeness to towns affects markets and opportunities for migration. Pastoral groups will generally need categories of their own. Often a classification based on general characteristics will be all that can be specified in briefs prepared for case studies in the early stages of the project, since no information about individual variations or characteristics will be available. Later, project records should provide a means of specific groupings; for example, adopters and nonadopters, those showing satisfactory gains and those with disappointing results.

Occasionally, community, social, or cultural factors may be very important, especially when the whole or most of the community must adopt the change for it to work. The maintenance of an irrigation system, dependent on community agreement over the disposal of water and the provision of local labor to keep it working, is an illustration. In instances such as this the community itself will be the focus of the enquiry.

The specific factors may be examined more fully by the characteristics in the following checklists:

(a) *Resources*
    (i) *Family and Labor*: size/age/responsiveness to farmers' directions/availability through year/other employment/labor hired
    (ii) *Land*: size/fertility/position (slope)/access to water/access to market/cropping pattern and timetable
    (iii) *Livestock*: type/size/meat/milk/capital
    (iv) *Capital*: Equipment owned/equipment hired/stores/vehicles (including bicycles and animal-drawn carts)

(b) *Attitudinal*
    (i) Age, education, and past experience
    (ii) Position in community
    (iii) Sensitivity to risk
    (iv) Influenced by educated or migrant sons or daughters.

Many of these variables listed will not be known before the study is undertaken, but will be included in the study as potential explanatory causes of response variation. Such individual information as is available, however, may be used, even at the planning stage, to provide further subgrouping.

The more groups that are formed, the greater the homogeneity within each group. But the complexity of organization of the study increases for which there has to be some tradeoff. The crucial point is that the variability *between* groups in relation to the topic under investigation should be substantially greater than the variation *within* the groups.

It should be noted that once the specific factors are brought into account, group boundaries will seldom be geographical; farmers in the same local community may be in different groups. However, in order to get the depth of insight promised by case study methods, the investigations must be limited to a few sites, and to a small number of farmers.

The case study should not be an isolated enquiry; the topic and the selection of relevant groups are guided by the place of individual studies within the total monitoring and evaluation effort. Its planning takes place against the background of the monitoring series, and its detailed setting-out is sharpened by special, preliminary, rapid observation surveys. Its findings are interpreted alongside any sample surveys planned jointly with the case studies. With sample surveys running in parallel the point made earlier is strengthened; namely, that the case study will not necessarily be looking at the 'average'. Greater insights may sometimes be expected from a study of the extremes—for example, those for whom the project has worked well and those who have rejected it entirely.

PART 7:
SAMPLING AND SAMPLE SURVEY DESIGN FOR MONITORING AND EVALUATION

7.1   The Basic Logic of Sampling

If data are collected from every member of a population—a census—no inference is required to relate the results to this population; the only sources of error are those involved in the enumeration and the analysis, such as faulty recording, refusal to respond, errors in responses, omissions (or duplications), and erroneous coding and processing. These are nonsampling errors.

Censuses are usually impractical, except for simple counts of the population by national agencies, so recourse is made to sampling—the collection of data from a few members of the population, but with the intent to make inferences about the population at large. Another error is then introduced, namely, the sampling error, which arises because the sample chosen does not perfectly represent the population; the result obtained from drawing another sample from the same population would be different. But this extra error may be more than offset by a reduction in the nonsampling errors indicated above, due to the higher quality of enumeration that can be achieved by reducing the scale of the enquiry. The sampling error can be estimated from the sample data itself if each member of the population has a known probability of being included in the sample.

The logic of sampling may be expressed by considering a series of samples of certain size drawn from a population, with the intent of calculating the mean of a certain attribute of the population. Some of the samples will result in a mean below the mean of the total population, others will result in a mean that is higher and some will produce a result very close to the population mean. The important theorem is that the distribution of these sample means, if the samples are of reasonable size, will be close to that of a normal curve, which itself has a mean equal to that of the population. Because of this, the likely margin by which a given sample mean differs from the population mean can be calculated and is (with some reservations to be mentioned later) inversely proportional to the square root of the sample size.

For these results to apply, the principle of randomization must be introduced into the selection process. The population to be sampled may be grouped into various categories with different selection probabilities for each group; but, when this and other refinements of the design have been applied, the selection of one or more units from a group must be made at random. It is this essential element of randomization that distinguishes probability sampling, which allows inferences to be made with precise levels of confidence, from other forms of sampling, which, although useful in certain contexts, cannot provide proper estimates of sample error.

It follows, therefore, that in order to assign each member of the population a known probability of selection in a sampling process, the population must be fully defined—a frame i.e., a complete listing of its constituent units is required.

7.2 The Choice of Unit and the Sampling Frame

The selection of the ultimate sample unit is an important early design decision. Possible units include

(a) an individual person;
(b) a household;
(c) a cooperative or operational group;
(d) a village or social group;
(e) a plot;
(f) a holding (farm);
(g) a mapped area;
(h) a place where a specific activity occurs; and
(i) an area administered as a single entity.

Individuals as sample units are appropriate in demographic and certain attitudinal surveys; they are less appropriate in surveys dealing with economic and quality of life measures, where the economic or social unit is more relevant. Moreover, individuals are difficult units to list and keep track of, compared to households or farms. Surveys aimed at studying a specific activity may use as the sample unit the locations where this activity takes place, e.g., a market or clinic.

The selected unit must be precisely defined, otherwise, at the frame preparation, listing, and sample unit identification stages, ambiguities will arise, with different interpretations being made by different members of the survey force. As stated in Part 4, there are nearly as many definitions of a household as there are household surveys. A review of the concepts and definitions in local use should be made with a view to adopting the standard definition—if it exists.

70

The paramount requirements are that the definition must be unambiguous and also acceptable as rational by both enumerator and respondent. It is little use having a clear definition if it is seen as contradicting actual local custom.

Area sampling is particularly appropriate in rural areas where a map of the overall area may be the only frame available; this can be regarded as a set of small areas which can be sampled. However, the small areas must also be carefully defined and accurately mapped; unidentifiable boundaries on maps have been, and continue to be, one of the most serious sources of bias in sample selection.

If the population to be surveyed is a particular subset of the general population, such as project beneficiaries, administrative records listing this sub-set may exist. Such a frame should, however, be checked for:

(a) errors of recording;
(b) omissions; and
(c) duplications.

Usually a complete list of the units in the population does not exist; a sampling frame has then to be constructed in stages, giving rise to multi-stage sampling (see Section 7.6 below). Lists of administrative areas (too large to be considered as ultimate sample units), can be used to draw a primary sample of these areas. A further listing exercise is then undertaken to construct a frame of the required second-stage sample units *within the selected primary units*. If necessary, a third or fourth stage can be introduced with the frame being constructed at each stage only within the units selected at the previous stage.

These decisions on the choice of sample unit and the frame to be used for selection of the units are often hastily made, in order to commence the survey quickly. Biases are introduced, in consequence, that wreck the chance of achieving the objectives of the survey, no matter what the sample size and the care taken in data collection. In countries without cadastral mapping, samples of small areas using census enumeration maps have commonly been found (later) to produce estimates that are subject to biases in the 15-25% range.

The resources to be spent in constructing a frame from which to draw the sample depend, to some extent, on whether it is to be used for an ad hoc survey covering a single topic, or for a series of surveys over time, covering many topics. In the latter case, considerable effort is justified, whereas for the former there is a limit beyond which further resources to effect minor improvements in the frame are not justified.

7.3 Choice of Sampling Procedure

Some of the information required for monitoring and evaluation (particularly monitoring) may not require the use of sampling procedures. Records of inputs supplied, the beneficiaries of these inputs, construction work, and the like may be complete and readily summarized. Even when such complete records exist, if they are large in number and stored other than on computer files, sampling may be useful as a means of facilitating the regular (perhaps weekly), reporting of progress to project management.

To collect data on many of the indicators listed in Part 4, sampling will be necessary. A case study may be regarded as a type of sample; but as described in earlier Parts, the detailed microlevel investigation of a particular phenomenon may dictate that the cases are deliberately selected to reflect the phenomenon of interest. Any attempt to introduce a random selection in order to achieve sampling respectability may weaken what should be the strength of the case study—the detailed examination of carefully selected representative units. There are other sampling procedures, which, although failing to meet the strict randomization requirements, may nevertheless produce insights about the population at much less cost in time and resource than that required for random sampling. The following definitions identify the range of procedures:

> *A random (or probability) sample is the selection of a number of elements from a statistical population in such a way that every member of the population has a known probability of selection.*

A simple random sample is one where every element of the population has an equal probability of selection. Other designs that preserve the principle of randomization are discussed in Section 7.6 below. The precision of these other sampling designs is judged against a simple random as the standard, but this does not mean it is the most efficient sample to adopt.

> *A purposive sample is the selection of a number of elements of the population by a person exercising deliberate choice in an attempt to achieve a sample that represents the population.*

The intent to make numeric inferences about the population at large is still there (unlike the case study), but the selection is made on subjective grounds in order to minimize the costs of sample element identification or to maximize the convenience of the enumerator. It is hoped that the sample is representative enough for the purpose in hand, but this will not be known with any measurable degree of precision: the probability of bias is there, but its size is unknown.

> *A quota sample is the selection of a number of elements of the population in such a way that a given number of them fall within each of a defined set of subgroups of the population, but where the choice of the actual elements included in the sample for each subgroup is left to the discretion of the enumerator.*

The sample designer using purposive or quota procedures claims to be aware of the range of elements that make up the population and is attempting to cover them. This is a major improvement on the cruder rapid assessment techniques where observation is limited to the observer simply going along a road on his official rounds of visits to predetermined areas. Although these observations are indeed 'purposive' in one sense of the word, elements encountered using such a procedure are unlikely to include all types of interest within the population.

Depending on the circumstances in which monitoring and evaluation units are placed, purposive or quota sampling may be appropriate and acceptable. Consider the following examples:

(a) In an appraisal of an extension to the project area, it is required to assess quickly the range of farming patterns in the area, but no information exists regarding the size of the population and no frame of villages is available.

(b) A hypothesis that, within an existing project area, the reluctance to adopt the project package is associated with the land tenure status of the farmer has to be checked, but no records exist of the numbers by type of tenure.

(c) Yield differentials by size of farm need to be reviewed urgently, the proportions of farms within certain size groups are known from an earlier census, but no list of individual farms by size exists.

Suppose further that the survey has to be launched immediately with no time to establish a sample frame, and that resources are insufficient for the location of randomly selected segments.

For example (a), purposive sampling may be the only feasible solution. A few small areas can be selected, each of which has different climatic or soil conditions; the farming pattern for each selected area is ascertained by interviewing prominent local community contacts. These areas have been selected purposively with the intent to reflect the true range of farming patterns of the entire area; although there is no guarantee that this has been achieved, the result will almost certainly be stronger than would be obtained by interviewing centrally located officials whose claim to local knowledge may be questionable, or by using data from another area, or national estimates that are erroneously assumed to reflect conditions in the area of interest.

For example (b), a type of quota sample is possible. The types of land tenure may be known although the relative prevalence of each is not. The enumerators may be instructed to locate within their area of operations predetermined numbers of each type and to conduct an interview with each farmer. A rule must be established to cover the case where an enumerator cannot easily fill his quota for a particular type. After following some trace procedure (asking local authorities, etc.), he must be allowed to indicate that this particular type is rare or nonexistent in his area.

Example (c) allows for more precision in the quota sample selection. Because the relative size of each size-group is known, the quota for each size can be varied accordingly. Note, however, that there is no necessity to fix the quota in strict relation to the proportion of the population in the size-group. A size-group with a relatively small number of farms within it (e.g., farms greater than ten hectares in size), may nevertheless be of particular importance, and be given a quota equal to that of a size-group that contains a larger proportion of the farms. Again, the enumerator will be given the discretion of filling his quotas as best he can within a few general safeguard rules.

In each of these cases, the intention is to collect information from a sample, the results of which are to be taken to represent the population. Means, ranges, ratios, and proportions will be calculated and will be treated as representative of the value for the relevant group. For example (c), even estimates of population totals may be made as the sampling fractions are known. What cannot be done is to calculate the sampling errors of these estimates and thereby to calculate statistical confidence limits for them. With quota sampling, sampling errors are sometimes calculated on the assumption that the respondents were selected at random without selection bias, but this assumption is not usually justified.

This loss of power in the treatment of the results may not worry the user unduly. The vaguer caveats attached to the results, stressing the unknown size of the biases may be entirely acceptable—indeed, the recipient may be delighted to have at least this level of quantification of the problem. It may be noted that many reports of surveys involving probability samples are silent on the errors of any estimate, aggregated or disaggregated, contained in them. Numbers in individual cells of tables of cross-classified variables are treated as if they were as reliable as the marginal totals. Although such a practice is undesirable, it reflects the lack of demand by many users for such statements of precision, often resulting from unfamiliarity with this way of thinking.

The questions to be answered in choosing a sampling procedure may therefore be listed as follows:

(a) Has the existence of a frame or the possibility of constructing one been examined?

(b) Is other, more aggregate, information about the population and its distribution available and can it be used?

(c) If formal randomization is not possible, what rules and procedures can be adopted for the selection of the final sample units that will at least minimize the dangers of bias?

(d) Which design will serve to spread the sample as widely as possible, given the resources and time available?

(e) If nonrandom methods are indicated, have the users been made aware of the limitations and dangers involved in the inference processes?

The value of reported information concerning a population depends first on the accuracy with which the data are collected, as reflected in the skills, diligence, and care taken in the survey operations, and secondly, the accuracy with which inferences can be drawn from the data. Both depend on the sample design. One of these requirements should not be neglected at the design stage due to concentration on the other. The skill of the survey design is to achieve the optimum balance using the given limited resources. This is particularly true of surveys in rural areas where the inadequacy of prior knowledge and the difficulties in measuring key variables are not the exception, but the norm.

7.4   Required Sample Size and Survey Precision

The incorrect supposition that the required or desirable sample size is a percentage of the size of the population is still a surprisingly widely-held myth. Often, sample sizes are reported to be one percent, or five percent, of the population, as if this were a necessary statement to give the sample credibility. Or, a sample larger in one project area relative to that in another is justified merely because one area contains a larger population than the other. It is worth stating at the outset that:

If the total population of the survey area is very large, compared to the sample to be selected, the variance of, and hence the precision of, estimates calculated from the sample data is a function of the absolute number of sample units, not the sampling fraction.

The sample size required is a function of the variability of the characteristic measured, and of the degree of precision required. The variability component can be most conveniently expressed in this calculation in the form of the coefficient of variation (V), equal to the standard deviation of the estimate divided by the estimate (or σ/X).

The precision component requires two factors, D and K. D represents the largest acceptable difference between the value estimated from the sample and the true population value. It is also expressed in relative form. K is the measure of the confidence with which it can be stated that the result does lie within the range represented by $\pm$ D. The higher the value for K, the greater the degree of confidence. A value of K=2 is often chosen, since it provides a degree of confidence equal to odds of 19 to 1 (that is, a 95% confidence interval, as technical terminology puts it). A value of K=1 gives odds of 2:1.

Suppose that the standard deviation of a yield cut of a certain crop is equal to half the average yield of a cut. Then V=0.5. Suppose also that the user of the information will be satisfied if the sample estimate is within 10 percent of the true population value, that is, D=0.1. Then the required size of sample, n, is calculated as follows:

$$n = \frac{2^2 (.5)^2}{(.1)^2} = \frac{4 \times .25}{.01} = 100$$

What this calculation ensures is that when the average of the sample of 100 yield cuts, $\bar{x}$, and its sampling error (S.E.) are calculated, the result can be expressed in the following way:

odds of 19:1 can justifiably be taken that the
true value in the population lies within the range, $\bar{x} \pm 2$S.E.

Further, the range component, 2S.E., divided by $\bar{x}$ will be approximately equal to the acceptable difference, D, provided the prior estimate of V was reasonably accurate.

The calculation of n therefore requires three inputs about which the project management has to be consulted.

(a) The variability of the characteristic in the population. The user may have some idea of this from a previous study or 'general' knowledge.

(b) An indication of the acceptable error margin, D. For some enquiries, this can be relatively large; for others, it may be very small.

(c) The degree of confidence (odds ratio), with which he desires to be in this range.

The general formula then is:

$$n = \frac{K^2 V^2}{D^2}$$

Note that as K and/or V increase, n increases according to K^2 and/or V^2. Thus, if one characteristic has a coefficient of variation twice as large as another, n will have to be 4 times as large for the same level of precision. D^2 appears in the denominator. In the above example, D was equal to 0.1, D^2 to 0.01, and dividing by D^2 was equivalent to multiplying by 100. If D is halved to 0.05, D^2 becomes 0.0025, and dividing by D^2 is equivalent to multiplying by 400, that is 4 times the previous hundred. Thus, if D is halved, the sample size has to be 4 times as large. Reducing D by a factor of g requires that the sample size be increased by a factor of g^2.

As a further example, consider a project that covers a large state containing six zones or districts. Suppose it is desired to measure crop yields on the basis of yield cuts so that the estimate for each zone lies within 8% of the true value with 95% confidence, and the estimate for the state is required to be within 5% of the true value with 95% confidence. It may be assumed that the coefficient of variation of yield within a zone is 0.5, but at the state level it is rather higher, say 0.7.

Then the required sample at a zone level is given by:

$$n_z = \frac{2^2 (.5)^2}{(.08)^2} = \frac{1}{.0064} = 160 \text{ approx.}$$

whereas, at the state level:

$$n_s = \frac{2^2 (.7)^2}{(.05)^2} = \frac{4 \times .49}{.0025} = 780 \text{ approx.}$$

If the sample size within each of the 6 zones is 160, the required state level accuracy will be more than achieved.

Yet another example is provided by the need to estimate a proportion, p, of the population displaying a certain characteristic, e.g., the proportion adopting a project input. The sampling error of the estimate of p is approximately equal to $\sqrt{\frac{p(1-p)}{n}}$, p lying in the range 0 to 1. Thus, given a required sampling error and an assumed value for p, the required sample size can be calculated.

If p is expected to be of the order of 0.5 and the sampling error acceptable is 0.05 then

$$n = \frac{0.5 \times 0.5}{(0.05)^2} = 100$$

Since the numerator of the estimate of the sampling error is p (1-p), the absolute error for p = 0.95 is the same as the error estimated for p = 0.05 for the same sample size. The maximum value of the absolute error occurs when p = 0.5 and it does not vary much over a range of 0.25 - 0.75 for p. The relative error (or coefficient of variation) is, of course, much higher for the p = 0.05 compared to p = 0.95 even though the absolute error is the same. It is this relative error that is often of most practical importance.

Several points need to be made about the above sample size examples.

(a) Simple random sampling has been assumed. Further gains in efficiency can be achieved by sample design techniques such as stratification, but loss of efficiency may be involved in other design features such as clustering (see below).

(b) The sample size may be modest for a given estimate, for a given geographical area. But if estimates are required for individual cells of a cross-classification of two or more variables, or are required for subareas within the primary area, the total sample sizes to maintain a given level of precision at these disaggregated levels are greater than those for the primary area. In general, when detailed cross-classifications of any kind are involved, the number of cells in the tables rapidly becomes large, and the entries in them smaller. A rough guide to the sample size required can be calculated as follows, when the entries in the tables are counts.

An approximate estimate of the sampling error of the count appearing in a cell in a table is equal to its square root. Suppose for all groups forming five or more percent of the population, it is desired that their estimate should have a maximum relative sampling error (V, as previously described) of 0.2. Then, if the count of units in the sample in the cell is labelled Y, the requirement is that:

$$\frac{\sqrt{Y}}{Y} < 0.2$$

Or, Y must be at least 25. If this group forms 5 percent of the population, it will also form about 5 percent of the sample. The sample size must therefore be:

$$\frac{100}{5} \times 25 = 500.$$

(c) The user specifies a certain required precision in the estimate. In the size calculation set out above, so far the assumption is that the only error involved is the sampling error. This assumption will not hold in practice, and the implications are examined in more detail below (7.5).

Much of the monitoring and evaluation analysis will involve the significance of differences observed at two or more time intervals. Measuring a variable at different time-points raises a number of issues.

(a) How frequently should the measurements be made?
(b) Should the same sample units be retained over time?
(c) Can the past data be used to improve the current estimate?

A full discussion of these issues requires a detailed exposition outside the scope of this Handbook, and only a few general pointers can be offered.

It is probably overambitious to attempt to measure proportionately small changes from one season or one year to the next. Exogeneous factors, such as climate, cause considerable short-term fluctuations that will disguise for some time a project-influenced trend. A time-series comprising a pre-project baseline, a repeat measurement during project implementation, a project completion measurement, and a post-project completion measurement, is the best basis that is likely to be available.

The decision regarding the retention of identical sample units over time depends partly on the practical difficulties of maintaining a constant sample. It also depends on whether the desire is to maximize the accuracy of the later estimate or the accuracy of the estimates of change over time. The optimal procedures for these two objectives differ; but, in most project information systems, the emphasis will be on the second.

If two sets of observations are uncorrelated, the variance of their differences is equal to the sum of the two variances. The same rule applies to a comparison of sample means. If the sets of observations are positively correlated, the sampling error of an estimate of difference is reduced by a factor involving (1-r), where r is the correlation coefficient. The larger the r, the greater the reduction. When measuring change, therefore, it is desirable to choose a procedure which maximises this reduction. This usually results in recommending that the same units be retained, since measurements of the same variable in the same units at various points of time are usually highly correlated.

It may be possible to maintain a high proportion of farmers in a sample over the project period to measure change. But if the survey involves a high frequency of visits to the respondents, the constant interaction between the enumerator and the respondent may change the behaviour of the respondent so that he no longer remains typical of the population that, as a sample unit, he is supposed to represent. Moreover, the question of imposing excessively on the goodwill of an individual over a long and continuous time period must be taken into account. (See Part 3 for further discussion of these points.) For a combination of these reasons, there will normally be a need to rotate or replace the membership of the original sample, although the primary *area* sample units of a multistage sample design should be retained for as long as feasible. This rotation of the sample enables a chain linkage to be established, as the sample for any two consecutive periods will contain a proportion of common units. The impact of new units is, as it were, spread out over the extended time period, reducing the disturbance in the time-related trend line that is the main paratter of importance. As a general rule, a replacement or rotation of about one-quarter to one-third of the sample at each round seems to work reasonably well.

In addition to such a planned rotation there will be an inevitable attrition rate due to death, migration, or a change in household composition leading to the splitting or aggregation of units.

7.5 Components of Survey Error

The statements about the precision of estimates assume that the observed values are accurate—the only error to be considered is the sampling error. Because of this, there is a tendency in reporting survey results to quote error margins (if indeed they are quoted at all) in these terms. However, the measurement of variables such as production or income of semi-subsistence agriculturalists creates serious interviewing and measurement problems as discussed in Part 4.

Even objective measurements such as yield cutting may be subject to substantial errors unless the techniques used are well designed and rigidly controlled. Response errors to questions arise from reluctance to reveal the true answer, incomprehension of the question, ignorance of the answer in quantitative terms, wish to please by giving an answer that is assumed to be favorable, inaccuracies due to faulty memory, and many other factors. Unlike the sampling error, these errors are not random in nature. They usually tend to bias the estimate in one direction and it is singularly fortunate if two components of these errors act in different directions and cancel each other out.

The only prudent course is to expect substantial biases that may well exceed sampling errors.

The effect is summarized by the following equation:[8]

Mean Square Error = Sampling Variance + Square of the Bias.

When a user sets an error margin for the estimates from a sample, it is the root mean square error (RMSE) that is being thought of. Consideration of the possible relationship between sampling variance, related to the size of the sample, and the size of the bias helps in checking how RMSE may be minimized. Experience in measuring rural indicators has demonstrated that the biases can be controlled if great care is taken in training and supervising the enumerator force. This tight control over quality usually only occurs when the field force is of modest size. A small field force, however, limits the size of the sample that can be enumerated. Increasing the sample size beyond a critical level may mean that control over quality suffers, thus increasing the non-sampling errors. This increase may more than counterbalance the reduction in the sampling error due to the larger sample size. Unfortunately, the biases in-

---

8. The proof of which can be found in advanced statistical texts, but is not given here.

volved for different sizes of survey organizations cannot easily be foreseen; the optimal balance between sample size and data quality cannot be precisely determined. However, experience suggests some general guidance.

An example in Part 7.4 gave a result that a sample of 780 would give a maximum difference of 5%, implying a sampling error of approximately 2.5%. If each enumerator can interview 20 respondents, an enumerator force of 39, supplemented by approximately 8 supervisors, is required. If this sample was doubled to 1560, the sampling error would be reduced from 2.5% to about 1.8%, but the field force required would rise to 78 and the supervisory force to 16. But if, as is possible, a skilled supervisory force of this number is not available, the resulting loss of control over quality could lead to increases in the biases which could result in an overall increase in the RMSE. Doubling the sample size may, therefore, result in a less accurate estimate. It should be noted that doubling the sample size does not always require a doubling of the field force. It may be possible to extend the duration of the survey.

Results of crop-cutting in India show that if they are carefully carried out by skilled enumerators, using an adequate size of crop cut, the bias in the resulting estimate can be reduced below 5%. However, badly conducted yield cuts give overestimates of yield in the 10-15% range—somtimes, higher. Suppose that the monitoring and evaluation unit can only field a skilled force sufficient to deal with 200 yield cuts, and that the standard error of this sample is 10% with a bias of 4%, then the RMSE is given by:

$$RMSE = \sqrt{10^2 + 4^2}$$
$$= 10.8\% \text{ approx.}$$

If the standard error was to be reduced to 5%, the sample size would need to be increased to 800. But, at this level, quality control is lost, and the biases rise to, say, 10%. Then,

$$RMSE = \sqrt{5^2 + 10^2}$$
$$= 11.2\% \text{ approx.}$$

Quadrupling the sample has reduced the accuracy of the estimate.

It is proposed, therefore, that when dealing with variables subject to large response or observational biases, which can only be reduced by in-depth interviews or carefully supervised measurements, the following slogan with regard to sample sizes is appropriate:

Small may be better.

The effect of bias on estimating changes over time is rather different. If the bias is constant over time, the RMSE of the *difference* between two observations at different time periods will be relatively unaffected by it. In such a case, a biased estimate from a large sample, repeated over time, might be more effective in detecting change than an unbiased estimate from a small sample with a larger standard error. However, as discussed in Part 4, the bias may not be constant over time.

7.6　Useful Sample Design Options

So far, the discussion of sampling error has been presented in the context of simple, random sampling. There are other options open to the sample designer, given a frame from which to sample, which adapt the principles. Some of these options improve the precision of estimates for a given sample size; others make the survey operation more practicable, at the cost of increasing the sampling error. Refining the sample design can be taken too far. In the rural development context the frame is likely to be inadequate, the cartographic experience limited, and the enumerators inexperienced. The more complex the design, the more likely it is that large biases will be introduced due to errors in sample identification. Nevertheless, within limits, the following features should be considered:

(a) stratification;
(b) systematic sampling;
(c) multistage and cluster sampling;
(d) self-weighting samples; and
(e) interpenetrating samples.

(a)　Stratification

Information available for each unit in a frame can be used to allocate the units into groups according to a characteristic, so that each group of units is more homogeneous than the population. If a separate sample is then drawn from each group, the result is a stratified sample, each group forming one stratum. Thus, in the example given in Part 7.4 of six zones within a state, the zones might be treated as strata, from each of which a sample is drawn.

Estimates of averages, totals, etc., from the sample are derived from the individual stratum estimates. The variance of the overall estimate is the sum of the individual stratum variances: the variance *between* strata does not contribute to the sampling error. If, therefore, different strata of relatively homogeneous units can be formed, the precision of the overall estimate for a given sample size may be substantially improved. Or, alternatively, the same level of precision as that of a simple random sample can be obtained using a smaller sample.

This arises in the following way:

Let x_{ij} be the jth observation in the ith stratum.
$\bar{x}_{i.}$ = the mean of the x_{ij} in the ith stratum.
$\bar{x}_{..}$ = the mean of x_{ij} over all strata.

The unstratified variance may be seen as a multiple of $\Sigma\Sigma(x_{ij} - \overline{x}_{..})^2$

The stratified variance may be seen as a multiple of $\Sigma\Sigma(x_{ij} - \overline{x}_{i.})^2$

Now $\Sigma\Sigma(x_{ij} - \overline{x}_{..})^2 = \Sigma\Sigma \ [(x_{ij} - \overline{x}_{i.}) + (\overline{x}_{i.} - \overline{x}_{..})]^2$

It can be seen that the expansion shown contains the stratified variance as but one component. The between-strata variance indicated by $(\overline{x}_{i.} - \overline{x}_{..})^2$ is contained in the unstratified variance but is not included in the stratified variance.

Points to note include

(a) all strata must be sampled otherwise no overall population estimate can be obtained;

(b) to calculate a variance there must be at least 2 units within each stratum; and

(c) little is gained if most strata are reasonably homogeneous, but at least one exhibits very wide variations; the contribution to the variance from the single 'weak link' may be as large as the sum of the others. This is why the open-ended catch-all stratum (see example below) usually requires a much higher sampling fraction.

Effective stratification is the most powerful tool for improving precision. There is very little risk attached to its adoption. Even if the stratification is not as successful in achieving homogeneous strata as was hoped, some benefit will almost certainly accrue. In any case, most stratifications in agriculture and rural development projects are likely to group the population in classes that are useful for study, quite apart from the increase in precision.

Information for stratifying the population should be related to the variables that are to be measured. If the survey is a multisubject one, stratification according to one characteristic may be very desirable for the estimation of certain variables, but may contribute little to improving the estimation of others. The criterion for stratification in such a case will be to improve the estimation of the most important variables. The commonest stratification is by geographical subdivisions of the project area. It may be expected that, as a matter of course, units in the same geographical area will exhibit some degree of homogeneity.

Administrative records may indicate the recipients of, say, credit, by the amount received. A sample survey into the use of credit would clearly benefit by stratification using this information. Surveys of farms would gain much from stratification by type or size of farm, if,—and unfortunately it is a big if—the frame contains this information. Stratification by soil type is also an attractive possibility, but if this involves grouping subareas across existing administrative boundaries, complications may ensue in controlling the field operations.

It is not necessary to maintain a constant sampling fraction for each stratum. Calculating the optimum sample allocation to strata requires knowledge of the variances within each stratum: these are often not known exactly, but a pattern of relative variances can be assumed according to general information.

Consider the following example of strata formed by farm size:

Table 2: Number of Farms by Farm Size

STRATUM	AREA (Hectares)					
	- 1	1 - 3	3 - 5	5 - 10	10 - 50	50 +
NO. OF FARMS	5,000	7,000	2,000	700	200	100

Of the 15,000 farms, suppose it is desired to select 600. A fixed 4 percent sample across all strata will give only 8 and 4 selected farms in the two-strata containing the largest-size farms. But these strata, particularly the open-ended last one, will exhibit large variations in farm size, production, etc. A more efficient sample may be obtained by selecting 100 from each stratum. This means that the sampling fraction varies from 1/70 in the second stratum, to 1 in the last stratum, i.e., all large farms are included.

This may be an efficient sample design in one way, but the benefits of reduction in error have to be traded off against the greater complication of the calculation of estimates and their variances. The sample is not self-weighting across strata; i.e., the probabilities of selection are not equal for all units in the population. (See the section on self-weighting samples below.) The optimum allocation of sampling fractions for each stratum depends, of course, on the relative variance within each stratum. Even a pragmatic allocation, such as the example given, may be very beneficial.

The proportions of the population within a number of possible strata may be known, but a frame of individual units with the required information to allocate them to the correct stratum may not be available. Post-stratification, whereby the sample units are allocated to strata after the data have been collected, may be useful. Note that the weights to be given to each stratum need to be known. As long as the sample number in each stratum is a reasonable size (more than 30 is an approximate guide), this procedure will achieve some of the benefits of ordinary stratification.

(b)   Systematic Sampling

If the population units are contained within a limited area, e.g., farmers below a small dam, all of which is easily accessible at minimal cost, a sample of respondents may be selected directly with no intervening stage. Sampling of administrative records is a particular case of single-stage sampling. If the records are contained in files and any individual record can be easily located, the sample can be selected *systematically* from the files.

Consider, as an example, the card index of a credit bank containing the names of credit recipients and the amount of credit issued. The review of a sample of recipient records will often be sufficient to monitor the distribution of credit by recipient. Systematic selection adopts the following procedure. If 500 records are to be sampled out of 5,000, a random number between 1 and 10 is chosen to determine the first sample unit. The remaining units are selected by adding 10 to the initial random number.

In the general case:

Calculate $\frac{N}{n}$ where N = total number of records.
$\qquad\qquad\qquad\quad n$ = number required in sample.

Select a random number r between 1 and n
The sample units are then r, $r + \frac{N}{n}$, $r + \frac{2N}{n}$ . . . . , $r + \frac{(n-1)N}{n}$.

A systematic sample can be regarded as equivalent to a simple random sample if the records are stored in a sequence that is unrelated to the variable to be measured; e.g., filing in alphabetical order by beneficiary (it is unlikely that the amount of credit is determined by the starting letter of a beneficiary's name). If, however, the file is maintained in a sequence related to the size of credit, systematic sampling will be better than simple random sampling and approach a form of stratified sampling. The main risk of systematic sampling is that the file of records contains a periodicity that is in some way correlated with the sampling interval; in such a case, a systematic sample will be biased. The danger is small in most practical applications, but must be checked.

(c) Multistage and Cluster Sampling

Often the only frame available for selecting households or individual beneficiaries is a record of the *number* of these units in each of the project subareas. A two-stage sample will then be appropriate. The first-stage sample unit (primary unit) is the project subarea. Having selected a number of these, a list of the relevant individual units is compiled locally within each of the selected primary units. A second-stage sample of the final sample units is selected from each of these lists.

More stages may be introduced. The primary unit may be a subdistrict, the secondary unit an area block within selected subdistricts, the third stage fields within the selected blocks, and the final stage small segments within the selected fields for yield-cutting purposes. However, each extra stage, with local intervention required to identify and list the units in order to proceed to the next stage, increases the risk that something will go awry.

The use of geographical areas as primary units is termed cluster sampling. This, together with stratification, is the most common feature of a sample design.

A clustered sample bestows certain benefits—final units have to be listed only in selected clusters and a sample of secondary sample units (e.g., households) within a cluster will be grouped closer together in terms of physical location than a simple random sample of units from the population. This reduces the cost of identification of the sample units and the travelling of the enumerator moving from one respondent to the next. However, units within a cluster may be similar with regard to characteristics of interest, so that the contribution, in terms of improving the precision of the estimate, of including an extra sample unit within the cluster is much less than that which would be made by an extra unit outside the cluster. Thus a sample of n units made up of a set of m_i units in i clusters is less efficient than a simple sample of n units dispersed randomly in the population. One could, in theory, construct clusters that contain heterogeneous units, but these would not necessarily be single compact geographical areas, thus negating the main reason for adopting this approach.

To consider the cost of cluster sampling in terms of loss of efficiency for a given sample size it is necessary to introduce the concept of intraclass correlation. This measures the extent to which units within a cluster are more alike than units in different clusters. It is defined as:

$$\delta = \frac{\sigma^2_b - (\sigma^2/M)}{(M-1)(\sigma^2/M)}$$

where σ^2_b = variance between clusters
σ^2 = total variance
M = number of units within a cluster.

If M is large an approximation to δ is given by σ^2_b/σ^2, that is, the ratio of the between-cluster variance to the total variance.

The relative efficiency of a simple random sample compared to a cluster sample is given by:

$$z = 1 + \delta (\bar{m} - 1)$$

where $\bar{m}$ = the average number of units *selected* within a cluster.

It is clear from this expression that $\bar{m}$ will need to be kept small if δ is positive and large; otherwise the loss of efficiency becomes serious.

For example, if $\delta = 0.1$ and $\bar{m} = 20$

$$z = 2.9$$

i.e., a cluster sample would need to be nearly three times as large as a simple random sample in order to give the same precision.

The difference between stratification and clustering now becomes clear. With a stratified sample the between-strata variance is excluded from the sampling error calculation,[9] whereas with clustered samples the between-cluster variance can be a major contributor to the overall variance. Thus with stratification the aim is to maximize the homogeneity within-strata, but with cluster sampling homogeneity within clusters reduces the efficiency of the sample. When a stratified and clustered sample design is chosen, the gains in preci-

---

9. The proof of this can be found in advanced statistical texts, but is not given here.

sion achieved by stratification are, as it were, dissipated by the loss in precision caused by clustering. A reasonable compromise may be to aim for a net result roughly equivalent to that of a simple random sample of the same size.

Section 7.4 demonstrates that the required sample sizes for moderate precision levels need not be very large, however large the project area. The damaging effect of clustering is that, depending on the variable measured, a sample several times larger than those shown in Section 7.4 may be required for the same level of precision.

In advance of the survey, it is unlikely that δ is known. With the safe assumption that for many agricultural variables it will be positive, a rule of thumb is that m should be kept within single figures.

A clustered sample design requires a listing of all units within selected clusters in order to select the second-stage sample units. In the case of an urgent demand for one or two estimates of simple attributes of a target population, e.g., the adoption rate of a given technical innovation for the current season, it may be cost-effective to skip the separate listing stage, and briefly interview all the units in the cluster.

The number of units selected within a cluster is also higher than that indicated above, when an enumerator is stationed inside a cluster and it is wished to provide a full work-load.

In the case of a survey of access to essential amenities, the intraclass correlation may be so high that little is gained from extra interviews above a minimum of 3-6 per cluster. At one extreme there are certain questions that can be asked of a single community leader and the replies will describe the entire cluster, e.g., number of clinics, schools, water-holes, etc. It would be nonsensical to ask these questions of a number of individuals—but this has been known to happen.

A common form of stratified cluster sample in agricultural enquiries is as follows. The region is stratified into rational administrative and agricultural areas. For example, a project area could be divided into major administrative areas, with further subdivisions as necessary according to farming systems. This could give rise to 10-20 strata. For each of these strata a list of delimited clusters must be obtained—these could be census enumerator areas or villages. The lists for each stratum must be complete, and individual clusters must be identifiable.

A sample of these clusters is chosen within each stratum. As already mentioned, the sampling scheme for each stratum can be a simple random scheme with the same sampling fraction across strata, or it can be more complicated. A sample of clusters must, however, be selected from each and every stratum.

Only the selected clusters are included in the sample. They are identified in the field, and the final sampling units selected by one of the methods indicated earlier.

(d) Self-Weighting Samples

If every unit in the population has an equal chance of selection in the sample, the sample estimate is rated up to the total population figure by multiplying it by the reciprocal of the sampling fraction. If primary sample units, with unequal numbers of population units within them are selected at the first-stage, this self-weighting constant sampling fraction can be achieved if the probabilities of selection at each stage are chosen with this in mind.

One popular method is to select the primary units with probability proportional to their size (as defined in terms of population units), and then to select the second-stage units from each selected primary unit at a rate proportional to the reciprocal of the size of the primary unit. In this way the probability of selection of each unit is the same, no matter which primary unit it lies in—self-weighting has been achieved.

Thus, simplicity in the processing of the sample data in terms of weighting is retained. If electronic data processing facilities are available, assigning individual weights to groups of sample units will not significantly affect the processing speed; but simplicity should not be foresaken lightly. Self-weighting is particularly useful when detailed further analysis is proposed.

However, there may be practical problems. The allocation of probabilities of selection may turn out to have been based on very inaccurate prior information. Data may not be obtained from all selected sample units, and so 'missing data' techniques have to be introduced at the processing stage. Moreover, there may be other good reasons for varying the sampling fraction (particularly between strata, as discussed earlier).

(e) Interpenetrating Samples

If the selected sample units are viewed as two or more independent sub samples and the enumerators individually assigned particular subsamples, the analysis of error is both simpler and more powerful, allowing for estimates of the component of variability due to inter-enumerator differences. However, the practical implementation of these arrangements create logistic difficulties. The numbers involved may result in half the sample units in one cluster being in one subsample, and the other half in the other, with two enumerators operating in each cluster. The administrative convenience and reduced logistic costs that are achieved by a clustered sample with one enumerator assigned to each cluster is sacrificed in order to achieve an appropriate analysis of variance. In the project evaluation context, and in the light of recommendations in this Handbook for trading off the demands of practicality against statistical rigor, such an elegant technique may be queried. If serious enumerator/respondent biases are suspected, however, the possibility of estimating inter-enumerator variability should be considered.

7.7   Conclusions

Sample survey costs are a function of the number of primary and final sample units and the volume of data collected per sample unit. A doubling of a simple random sample will improve the precision by approximately 40 percent; but because the field enumeration and logistics probably account for one-half or more of total survey costs, and because increased volumes of data incur extra processing costs and time, the extra cost may be greater than the value of their gain in precision. Moreover, greater precision due to a larger sample size may be more than offset by an accompanying increase in bias. Thus larger samples costing more money and taking longer to analyze may produce little or no improvement in the overall precision of the estimates.

PART 8:
DATA COLLECTION BY INTERVIEW AND MEASUREMENT

8.1   Constraints in Interview and Measurement

In the preceding parts, various methods of data collection have been introduced, ranging from rapid subjective assessment to large-scale sample surveys. With the exception of the use of administrative records, all these methods involve the taking of counts and measures and/or the interviewing of selected respondents. Choice of the appropriate level of technique is important; measurements can be taken at different levels of accuracy, interviews can be conducted in a variety of ways, ranging from a free-flowing conversation to a more inquisitorial question/answer session. The general principles that should guide the choice are as follows:

(a) The information gathering process should be tailored to the respondents, so that it is acceptable to them and keep demands on them as low as possible, given the scope of the enquiry;

(b) If junior enumerators are to be used, their task must be within their capabilities and designed to minimize the difficulties they may encounter; and

(c) The data should be collected and recorded in such a manner that the subsequent analysis is facilitated.

Reports of sample surveys in developing countries reveal that the response rate to enquiries is very satisfactory; or, to put it another way, refusal to cooperate is rare, particularly in rural areas. This may seem to indicate that point (a) above has been well catered for; this is not necessarily the case. The small farmer in general, and project beneficiaries in particular, do not feel free to send an official packing even if the enquiry is offensive or arouses concern regarding the uses to which the replies will be put. Abuse of this passivity is not only wrong in terms of statistical ethics but also gives a false picture of the accuracy of the resulting, compliantly supplied data. Many of the enquiries in the project monitoring and evaluation context require the wholehearted cooperation of the respondent if the true situation is to be assessed; polite, indifferent or anxiety-motivated cooperation will not do. Prior notification of the purpose of the enquiry, minimizing the extent of 'sensitive' information covered, and limiting the demands on the respondents' time are major ways of gaining the respondents' full participation.

It is essential, particularly in sample surveys, that the respondents' time, or the access they provide to their property, should be used efficiently: the enumeration must be competently conducted. Too often, enumerators are burdened with a technique or a questionnaire that they cannot use efficiently. In this context, the survey designer must remember that a technique or question that seemed perfectly satisfactory in specially-controlled practice trials, may be more difficult to apply in an uncontrolled 'real-life' situation. The enumerator may not be able to obtain 'privacy' for the interview; the plots to be measured may be scattered over a very wide area; it may not be possible to interrupt wholesalers as they arrive at the market in order to question them on volumes and prices. Attempts to word questions so that they are unambiguous may lead to the use of technical expressions that are unknown to the respondents. The enumerator then has to 'explain' the question, which in practice often means that the respondent is led to an 'expected' answer.

Protecting the sensitivities of the respondent and facilitating the task of the enumerator are paramount considerations, but there is also the need to record the data in a manner that will lead to efficient and cost-effective analysis. Fortunately, the means of meeting this objective are usually compatible with the others. A clear, simple set of 'closed' questions are likely to be more acceptable to the respondent, more efficiently applied by the enumerator, and easier to code and process. A closed question is one where the range of possible answers can be listed, and then usually precoded.

The decision to use direct measurement/observation techniques or interviews depends on the background situation (see Part 4). Interviews are of little value when the respondent never knew the required information in quantitive terms, or, having known it once, cannot recall the figure at the time of the interview with sufficient accuracy.

A decision that such measurements are unavoidable has major implications for the scale and content of the resulting survey. They are demanding in skill and time and complicate the phasing of the enquiry. In small-scale enquiries, a whole package of observations and measurements can be included, as the enumerator will be spending considerable time with each unit of enquiry. The farm management study, with close observation of farming patterns and measurements of crop areas and yields, provides an excellent example.

Where appropriate information at an acceptable level of accuracy is to be collected by interview, the scope is constrained by the attitude of the respondent. If the interview is a long one, or the enquiry involves an excessive number of repeat visits, respondent fatigue and irritation become probable, interviewer efficiency is likely to decline, and there is a consequent loss of data quality.

8.2 Observation and Measurement

Common types of observation and measurement include

(a) instrument recording or readings, e.g., water flow or meteorological data;

(b) land and yield measurement, e.g., farm and plot areas and yield cuts;

(c) physical measurement or examination of individuals, e.g., health or nutritional status indicators;

(d) counts, e.g., of livestock numbers;

(e) direct observation, e.g., time to complete a certain task(period), or house structure/materials (instantaneous); and

(f) direct action, e.g., the actual purchase of market items in order to obtain prices.

Instrument recording where required is usually well-identified and understood.

Farm measurements are demanding in time, but since junior staff can be expected to master the necessary techniques (given adequate training and intensive supervision), they can be used in relatively large-scale investigations. Area measurements have been satisfactorily made using a prismatic compass and one of various means of measuring distances, such as range finders, chains, measuring wheels, etc. The important point to note is that the person taking the measurements should also have the ability and means to calculate the area on-site. A significant error rate on individual measurements is to be expected; and a single error, e.g., in a compass reading, can make nonsense of the result for the particular field. If these errors are only detected later, when the data is processed at some central location, considerable logistic difficulties ensue in arranging for the work to be redone, in some cases, it cannot be redone at all, as the situation has changed in the interim, e.g., the plot has been harvested. Thus, an on-site plotting of the bearings and the distances on squared paper in order to check that the diagram 'closes' appears imperative. Cheap pocket calculators which can be programmed to screen the data and calculate the resulting areas offers a quicker and more efficient method; the expense of such calculators will usually be more than justified in terms of man-hours saved.

Crop-cutting techniques use a small randomly located portion of the plot harvested by the enumerator. The size of the crop-cut may range from a few square meters to 100 square meters and the shapes commonly adopted are squares, circles, or triangles. Several points reflected in the literature should be noted.

(a) Even if measured accurately, the yields obtained represent biological yields rather than the actual yield obtained by the farmer. When harvesting entire plots, there will be losses that will not be incurred when dealing with small areas carefully harvested by an enumerator.

(b) The bias in crop-cutting, even after allowance is made for (a) above, tends to be towards overestimation of true yields; the size of the bias is approximately inversely proportional to the size of the crop-cut.

(c) Strict adherence to random location of the crop-cut is essential. Even well-trained enumerators tend to avoid bare or sparsely populated parts of the plot, even though the random coordinates result in the choice of such a portion. Such enumerator intervention will result in a serious bias in the overall estimate.

(d) Edge effects and border biases must be controlled. These reflect the tendency to include plants inside the crop-cut that are fractionally outside, and the tendency for the selection of crop-cut locations to under-represent the fringes of the plot (where yields tend to be lower). Once again, the net effect is usually towards over-estimation of true yields. This, in fact, may be the cause of the positive bias reflected in (b) above.

(e) Sampling only one crop-cut per plot or even per farm may be an efficient method of estimating an average yield for a zone or stratum. But the result of a single crop-cut says little about the average yield for that particular plot or farm. If plot or farm level estimates are required for individual farm budget studies, several crop-cuts per plot will be required. Due to the labor intensity of this method, several cuts per crop per farm make it difficult to operate a large sample of farms.

(f) Crop-cutting works best on evenly and densely planted plots of a single crop (irrigated rice being a prime example). Haphazardly sown plots containing two or more crops do not lend themselves easily to this technique.

The alternative of weighing the entire production should not be overlooked. Such a method may be feasible, especially for cash crops, which are not disposed of immediately or piecemeal. Supporting information for these types of crops may also be obtainable from the marketing agencies.

Physical examination of individuals is clearly a task for suitably qualified professionals or trained paramedical personnel. Data may be available from the records kept by such personnel at clinics, or as a result of case study investigations. An exception, where junior staff can be used, is the recording of anthropometric data (basically age, height, and weight), particularly of children as discussed in Part 4. Once again, the relationship between the level of measurement and the possible scale of the enquiry is clear.

Direct counts, whether of livestock, humans, tools, or sacks of produce, present only one major difficulty, namely, the bringing together of the items to be counted. This difficulty is very substantial, particularly for enquiries involving pastoral communities. The total cooperation of the persons involved is required; what is not revealed cannot be counted, and what cannot be counted at a few sites is unmanageable.

Direct observation of certain practices, particularly labor inputs, makes heavy demands on the time of enumerators. Great care is also needed to prevent the presence of the enumerator causing a change in the practice being observed. The respondent may work harder, spend more, or prepare a better meal (in a dietary survey), because the activity is under observation. A lead-in time, whereby the observations in the first visits are later discarded, may allow the respondent to settle down to more typical behaviour. This approach to data collection can be recommended only in microlevel studies.

In price collection, the value of direct purchases, as opposed to reported prices, is well documented. The decision, once again, depends on the required accuracy set against the size of possible distortions present in more casual responses. Price collection in rural markets is relatively simpler than the intractable problems of measuring crop production. But it requires careful definition of variety and quality; regular timing of observations; and due account taken of 'real', as opposed to, 'asking' prices.

One last word may be said on observation. Whatever the mode of enquiry, a keen eye can gather in casual observations impressions that provide an extra dimension to the results, even though these impressions are not quantifiable or suitable for inclusion in a survey questionnaire. One experienced practitioner, visiting beneficiary homes, noticed a high incidence of plastic tablecloths, where formerly there were few. This was a significant indicator of economic progress even though it had not been foreseen as an item to record. In short, subjective assessment by a trained observer may tell a story as important as that revealed by recorded data. In this context, the *absence* of an expected sign may be as revealing as its presence.

8.3   Interviewing: The Recall and Reference Periods

In any interview that has the objective of recording information relating to a specific time point or time period, the issues of recall and reference periods are important.

The recall period is the interval of time that has elapsed between the events that the respondent is asked to report and the moment of the interview. The reference period is the interval of time to which the question relates. Illustrations serve to clarify the distinctions between these two aspects:

(a) Interview in September, 1980 requesting details of planting at the beginning of the current season, say, March - May;
   recall period is 4-6 months;
   reference period is 3 months;
(b) Interview in December 1980 requesting details of crop sales in the agricultural year, April 1979 - March 1980;
   recall period is 9-21 months;
   reference period is 12 months.

Often the two periods coincide. For example, a question regarding births in 12 months preceding the interview has a recall and reference period of 12 months.

Detailed, rigid specifications may not be necessary in informal enquiries where the investigators are able to manipulate the framework of the discussion in a flexible fashion. However, a sample survey will require precise rules which must be applied consistently by all enumerators. A detailed discussion of recall and reference period issues is contained in many survey textbooks. A few general pointers are given here.

The reference period should be 'closed', i.e. the start and finish points should be clear, identifiable moments in the respondent's memory.

The difficulty is that what may appear to be a formally closed period may not be so in practice. This is particularly true of the use of calendar dates. A question may be phrased in terms of events in a particular year or month and this has precise starting and ending dates. But the respondent will not necessarily remember the required events that occurred in such calendar date terms. Events may be reported as occurring within the period which, in fact, took place outside it and vice versa. In contrast, a question such as "How many times did you weed between planting and harvesting of the season's crop?" may be 'closed' to the respondent as the weeding is bounded by two identifiable time periods that are linked to the activity itself: weeding cannot occur before planting or after harvesting. Of course, if planting was staggered over an extended time period the reference period becomes 'open'. If the local market operates on a single day in the week, a question concerning activities on the last market day is 'closed', unless there is ambiguity about the market under reference.

Memory 'fade' of most events is proportional to the time that has elapsed since they occurred. The more routine or insignificant the event, the shorter will be the period in which precise identification of it is possible. Major events will be remembered longer, although precise dates may be forgotten. The optimum recall period will, therefore, vary according to the topic under discussion. Expenditure surveys provide a good example. Major but uncommon expenditures (e.g., household equipment purchases) require a long recall period—perhaps one year. Expenditures on routine daily items such as food purchases will require a very short recall period of no more than a few days. Unfortunately, the shorter the recall period the higher will be the variance of the data (akin to the high variation in the results of crop yields estimated from small subplots). Moreover, the frequency of visits required, if very short recall periods are used, makes the survey expensive to execute. Improvement in the accuracy with which the data are collected by shortening the recall period must be balanced against the cost of doing so. Nevertheless, if the respondent is expected to recall events in detail, the recommendation must be to:

Choose a recall period that experience, or a pilot test, shows to be the maximum that provides a reasonable assurance that the events will be remembered.

A corollary of this may be formulated as:

Do not ask the respondent to recall something that was never known in the terms required.

8.4   The Type of Interview

The criteria for selecting enquiry procedures introduced in Part 5 included the type of interview. The following are the key issues:

a) Should the interview be unstructured and free-ranging or should it be structured and limited?
b) Should the questions be 'closed' or 'open-ended', (note that 'closed' here refers to type of question, not to reference periods)?
c) To what extent and in what circumstances can attitudinal perception and questions be included?

The interview that is conducted in the form of a flexible, but focused, dialogue, without formal questioning or concurrent detailed recording of answers, is a powerful survey tool when it is used by a skilled and professional interviewer closely involved in the enquiry as a whole. The possibilities of this type of interview can best be illustrated by an example; what follows draws on the experience of CIMMYT in planning technologies appropriate to farmers.[10]

The use is described of an 'exploratory survey' to gather information through informal interviews in order to understand farming practices and to ascertain farmers' problems. This survey is seen as an aid to designing a follow-up formal survey using a questionnaire to quantify the variables of importance as identified in the exploratory survey. It is also used to collect information that may be too sensitive or complicated to include in a formal survey.

"the interviews should be carried out in the farmer's field in order to relate questions to observations in the field. Interviews should be conducted in a relaxed manner. Use of pencil and paper should be avoided . . . however, the researchers should immediately note down all relevant information after leaving the farmer . . . It is possible to cover only a part of the information listed in our checklist in one interview with a farmer. What information is included will depend on what practices a farmer is following, what problems he is experiencing, and the degree of cooperation encountered. For example, a farmer who is experiencing difficulty in completing the . . . weeding desired (observed in the field visit) might be asked detailed questions about the hired labour market, competing labour demands . . . timing of operations, etc. . . . It is useful to ask some general questions about the crop and the farming system and then use the responses to decide what specific areas will be emphasized. It is not necessary to focus the questions on the practices of a specific farmer. In fact, much can be gained — particularly in interviews with traditional leaders — by discussing practices and variations commonly followed by farmers in the area. For these types of questions, interviews with groups of farmers can be particularly valuable in gaining rough estimates of the frequency of use of various practices . . . After each day's work, it is useful for the researchers to evaluate what they have learned, formulate new hypotheses and determine . . . the key gaps

---

10. D. Byerlee, M. Collinson et al., *Planning Technologies Appropriate to Farmers — Concepts and Procedures* (Mexico: CIMMYT, 1980).

and conflicts in their understanding which should be explored in further interviews."

The checklist referred to in the above extract is illustrated in Figure 3 below.

Figure 3: A Checklist of Information on Crop Management Practices

---

Land Preparation
  Sequence of operations
  Timing of each operation
    in relation to rains
  Equipment used in each
    operation
  Variation in method with
    seasonal conditions

Planting
  Variety(ies) used
  Density and spacing
  Density and spacing of
    interplanted crops
  Time of planting in relation
    to rains, frosts, etc.
  Spread of planting dates
  Sequence of interplanting
    crops
  Method of planting (hills,
    broadcast, etc.)
  Method of covering seed
  Practices of replanting part or
    whole fields

Thinning
  Timing
  Target density
  Use of thinnings

Weeding
  Number of weedings
  Timing of each in relation
    to planting
  Equipment used in weeding
  Use of herbicides (type, rate, timing
    and method of application
  Use of weeds

Fertilization
  Type of fertilizer(s) including organic
  Rate(s) of application
  Number and timing of applications
  Equipment used for application
  Method of application (e.g., broad-
    cast, furrows, etc.)

Pest Control
  Method of control (type, rate, equip-
    ment)
  Timing of control

Irrigation
  Method of irrigation
  Frequency and timing of irrigation

Harvest
  Timing of harvest in relation to
    maturity
  Method of harvesting
  Use of leaves and tops for animals
  Timing and method of picking leaves
    and tops
  Use of stalks

Post-Harvest
  Method of threshing/shelling
  Timing of threshing/shelling
  Method and quantity stored
  Disposal of produce
  Use of crop in local foods

Seed selection
  Time of selection
  Criteria for selection
  Special seed production or storage
    methods
  Seed treatment

The general approach is well illustrated by this example and the skills required of the interviewers are clearly highlighted. Investigators with these skills can probe attitudes and perceptions, and even explore potential reactions to hypothetical situations. However, great care must be exercised when using such questions for the following reasons:

(a) It is very easy to *lead* the respondent unconsciously to answers perceived as reasonable by the interviewer;

(b) Even if not led in this way, the respondent may assume that certain answers are expected and politeness results in meeting these perceived expectations;

(c) The use of technical or obscure words or phrases may be misunderstood, creating a situation where, either further explanation brings about (a) above, or the respondent is answering a different question to the one asked;

(d) If, to avoid (c), the question is put into what is believed to be simple, local parlance, it may then seem patronizing or, due to misunderstanding of the exact significance of local terms, misleading; and

(e) Questions requiring the respondent to place himself in a hypothetical situation are difficult to answer sensibly even by sophisticated respondents—the response may bear little relation to what will in fact be done, should the hypothetical situation become real.

The 'closed' attitudinal question, giving a list of optional answers of the 'tick one of the following' type, is considered under questionnaire design in Section 8.5 below.

Structured interviews using a questionnaire will, of necessity, cover a more limited range of topics, and is the required option when junior enumerators are employed. Within such a questionnaire, open-ended questions may be included in an attempt to obtain some of the advantages of a free-flowing interview. There are however advantages in using only 'closed' questions since:

(a) A neat questionnaire is easier to achieve (open-ended questions introduce uncertainties regarding the necessary size of the space that should be allowed for the answer);

(b) Precoding ensures that enumerators collect the information relating to the coding; and

(c) If the number of respondents is large, some form of coded data processing will be involved; open-ended answers will, therefore, need to be 'closed' at the coding stage, so why not earlier by the enumerator?

It is recommended that in a sample survey using junior enumerators, open-ended questions are used only to provide respondents with an opportunity to express their general reactions to the enquiry.

8.5. Questionnaire Design

Sample surveys demand a standardized format for recording the data; and case studies often benefit from this approach, at least for part of the work. The design of the questionnaire is, therefore, a matter of importance; failure to produce an appropriate set of questions, suitably sequenced in a practical layout, may jeopardize the success of the data collection before it is started.

In the monitoring and evaluation of agriculture and rural development projects, questionnaires are commonly needed to record data on the following topics:

(a) household, demographic and economic characteristics;
(b) farming patterns and farm economics;
(c) beneficiary attitudes and perceptions; and
(d) social information on beneficiaries and their communities.

The types of questionnaire applicable for these various topics vary widely and, as already indicated, the layout will depend on the level of knowledge of the respondent, the experience of the of enumerator, and the means of processing to be used.

The *content* of the questionnaire should be determined on the principle of including the minimum number of questions (or column headings) needed to meet the clearly specified objectives. The almost inevitable tendency is to ask more questions than is necessary. As discussed in Part 4, certain background information will be required to assist in the interpreta tion of the values of the main variables of interest, but the designer should have in mind a hypothesis of this explanatory role before including an item that is not of direct concern. The interview should be focused on the items of direct and major interest, not on secondary issues.

The *time* for the complete interview must be kept within bounds. Except in special circumstances involving an in-depth interview with an enthusiastically cooperative respondent, the length of the interview should not exceed one hour. This clearly limits the number of questions to be included; and the more detailed and thought-provoking the question, the more limited must the number be.

The questionnaire must be easy to *use* both as an enumerator guide and as an instrument for recording answers. The enumerator is faced with an often difficult and delicate interview situation, the tools provided should at least facilitate the task. Careful thought must be given to the quality of paper, the size of the sheets used, the clarity of printing and presentation, and the spaces provided for recording the answers.

The questionnaire must be *self-contained* in that the identification of the respondent, the enumerator, the date of interview, etc., are included. In many instances, it is necessary to check back with the enumerator or respondent when inconsistencies, errors, or unusual information are detected.

The questionnaire should, to the extent possible, be precoded, so that a manual office coding is avoided. Digital and coded responses should also be aligned, where appropriate, to ease the task of the computer data entry operator.

Three basic types of questionnaire can be identified, namely:

(a) the tabular 'row and column' form, on which responses are recorded, but which does not specify the wording of the question;

(b) listing of questions in the general order in which they are put, but without specifying the precise wording and without precise instructions regarding branching or alternative channels of progress; and

(c) the verbatim listing of questions with precise instructions on alternatives for progression, depending on the answers to preceding questions.

The Row and Column Tabular Format

This type of questionnaire is particularly popular for two of the main topics covered in monitoring and evaluation surveys: namely, the listing of household members and the recording of farm data, either field by field, or crop by crop.

Normally, the 'questions' are implied in the column headings and the replies for a given person, field, or crop are contained in one row. (Instead of fields or crops, the row entries may also, of course, cover items such as livestock type or farming practices.)

Two examples of such layouts are shown as Figures 4 and 5, illustrating a household composition form and a farm data form. The advantages and disadvantages of this type of layout are indicated in these examples.

The advantages are

(a) one sheet or form contains the results of what may have been quite a lengthy sequence of questions;

(b) the impossibility of specifying in advance the number of cases to be covered is allowed for by providing a large number of rows, one for each 'case'; and

(c) the enumerator is allowed flexibility in the manner and sequence of the implicit questions (but this may be a disadvantage also).

The disadvantages are

(a) questions cannot be fully specified, they are only implicit in the column headings;

(b) the lack of sequence and structure may result in omissions, the enumerator misdirecting the respondent, or the enumerator misrecording (e.g., placing the answer for one cell of the table in another);

(c) the abbreviated column heading may disguise the failure of the designer to realize that the implicit question is in fact a set of questions, some of which may be impossible to answer.

FIGURE 4

AN EXAMPLE OF A HOUSEHOLD COMPOSITION FORM

LIST BELOW
First all household members, present or absent
Then any other persons who slept last Sunday night in household

Today's date: _____   Last Sunday's date: _____

Full name	Relationship	Sex M or F	Was he in this household		Age (Enter col. 6 or col. 7, NOT both)		Marital Status	Tribe	Place of origin PROVINCE only	Is father alive? YES or NO	Is mother alive? YES or NO	ASK ALL FEMALES AGED 14 OR OVER							
			Last 12 Months YES or NO	Last Sunday YES or NO	Com- pleted years	Months if under 2 years						Of all the children borne alive by you in your whole life ...						Of your most recent live birth ...		
												How many are now in this household?		How many are now living elsewhere?		How many have died?		Date of birth		Child still alive? YES or NO
												M	F	M	F	M	F	Year	Month	
(1)	(2)	(3)	(4)	(5)	(6)	(7)	(8)	(9)	(10)	(11)	(12)	(13)	(14)	(15)	(16)	(17)	(18)	(19)		(20)

SUPERVISOR'S CHECK

Form checked in office ☐
Interview watched ☐
Reinterviewed ☐

Supervisor's Signature _____
Date _____

Enumerator's Signature _____

FIGURE 5
CROP PATTERN BY PLOT

TO BE COMPLETED BY ENUMERATOR

For each plot owned or operated by the Household you should record the crops present on the plot at time of your interview. Do not record crops until they have been planted or after they have been harvested.

TO BE COMPLETED BY SENIOR ENUMERATOR

Reg. `1`

`12` `0 5`

WEEK	Field Name	Field/Plot No.	Crops present now	Cd No. 14 15	Plot No. 16 17 18	Crop Code 1 19 20	Crop Code 2 21 22	Crop Code 3 23 24 25	Crop Code 4 26 27 28	Crop Code 5 29 30 31 32	Crop Code 6 33 34 35 36 37	Crop Code 7 38 39	Total Crops 40 41	Skip	I.P. No. 74 75	Type 76 77 78	Next Cd No. 79 80
				0 1													
				0 2													
				0 3													
				0 4													
				0 5													
				0 6													
				0 7													
				0 8													
				0 9													
				1 0													
				1 1													
				1 2													
				1 3													
				1 4													
				1 5													
				1 6													
				1 7													
				1 8													
				1 9													
				2 0													
				2 1													
				2 2													
				2 3													
				2 4													
				2 5													

101

Sometimes an attempt is made to combine the tabular format with verbatim questions and coding instructions incorporated in the column headings. Figure 6 provides an example. There are differences of opinion on the usefulness of this approach. It has been criticized on the grounds that the space taken up by the headings reduces excessively the space available for recording the answers, and that the printing of questions in full results in column widths varying according to the length of question, rather than according to the demands of the answer.

Whether it is wise to allow the considerable flexibility to the enumerator that the tabular layout demands depends on the topic of the survey, the quality of the enumerator, and the type of respondent. Certainly, careful training and intensive supervision is particularly important if this format is adopted. Specifically, the enumerator must be totally familiar with the definitions and implicit questions as summarized in the column headings. Reference to a manual during an interview is not practical, so the necessary instructions must be in the enumerator's 'on-line memory', to borrow computer storage phraseology.

Making the column heading simple does not simplify the question. Thus, a column heading 'Income' on a form intended for small semisubsistence farmer respondents does not solve the problem of determining income. Nothing betrays an inadequate questionnaire designer more clearly than such an abrogation of responsibility whereby the load is placed where it does not belong—on the enumerator.

The Nonverbatim Questionnaire

A common type of questionnaire presents the questions by topic and follows a sequence that is intended to reflect a logical progression of the interview through the topics being investigated. However, the questions are not printed in full verbatim form and may not include the branching and alternative channels of progress through the interview. This format does not usually require the questionnaire to be printed in the language of the interview. Nevertheless, the question is phrased in approximately the form in which it is to be put, perhaps in a shorter form than that to be used during the actual interview. It is, therefore, an intermediate type between the tabular and the verbatim questionnaires, in which a measure of flexibility for the enumerator is retained. How the question will be phrased in translation, the follow-up that may be required in clarification, and the recognition that a later question has been rendered unnecessary by a previous reply, these are at the enumerator's discretion.

As with the tabular format, this can be either a strength or a weakness. The essential requirement is that the topics and questions are simple in concept and will be well understood by the respondent. The enumerator should not be

FIGURE 6
AN EXAMPLE OF A HOUSEHOLD QUESTIONNAIRE

ASK OF HEAD OF HOUSEHOLD OR OTHER ELIGIBLE RESPONDENT
ASK QUESTIONS COLUMN BY COLUMN

NO.	NAME	SEX	DATE OF BIRTH				AGE	MARITAL STATUS	RELATIONSHIP TO HEAD OF HOUSEHOLD		STATUS OF THE MOTHER
			MONTH		YEAR						
			WEST.	ISLAM/JAVA OTHER	WEST.	ISLAM/JAVA OTHER					

NAME — A1 What are the names of the members of this household?

WRITE IN THE NAMES OF HOUSEHOLD MEMBERS

AFTER WRITING ALL THE NAMES GIVEN BY R. READ THE NAMES AGAIN THAT YOU HAVE WRITTEN AND ASK:

Are there any other household members who have not yet been mentioned, such as:

1. Young children or new-born babies?
2. Household members who are temporarily away from home?
3. Other household members such as other relatives or servants who live with you?

IF YES, ENTER THEIR NAMES

SEX — A2 Are you

Male — L
Female — F

FILL IN CODE

DATE OF BIRTH — A3 In what month and year were you born?

FILL IN MONTH AND YEAR OF BIRTH IN THE WESTERN OR ISLAMIC CALENDARS OR OTHER.

AGE — A4 How old are you?

WRITE AGE IN YEARS

MARITAL STATUS — A5 Are you currently?

Single — 1
Married — 2
Widowed — 3
Divorced — 4
Separated — 5

FILL IN CODE

RELATIONSHIP TO HEAD OF HOUSEHOLD — A6 What is the name of the head of this household?

What is the relationship of each person with the head of this household?

WRITE IN THE RELATIONSHIP WITH THE HEAD OF HOUSEHOLD

FILL IN THE CODE SHOWING THE RELATIONSHIP TO THE HEAD OF HOUSEHOLD

Head — 1
Wife of head — 2
Own child of head — 3
Non-own child of head — 4
Grand child — 5
Parent of head — 6
Parent of wife of head — 7
Son/daughter-in-Law — 8
Other family — 9
Non-family such as servants, etc. — 0

STATUS OF THE MOTHER — A7 Does your mother live in this household or outside of this household, or is she already dead?

Lives in this household — 1
Lives outside this household — 2
Already dead — 3
Don't know — 4

FILL IN CODE

Column numbers: (1) (2) (3) (4) (5) (6) (7) (8) (9) (10) (11) (12)

forced into difficult translations of technical terms[11] or made to clarify the question in such a way as to 'lead' the respondent to the answer. How much discretion can be allowed needs careful consideration and the training requirements are as demanding as those for the tabular questionnaire as outlined above.

Given that the precise wording of the question is not a requirement, the designer has an easier task in producing a layout that meets the general principles laid out earlier, particularly the ease of use and ease of coding and analysis requirements. An example is shown as Figure 7.

This type of questionnaire is well-suited when using printed lists of optional answers with the instruction to the enumerator to indicate which answer is closest to the respondent's opinion. However, the preparation of the list of options requires detailed knowledge of the issue in question; it is very easy to impose a set of answers that have no real bearing on the true reaction of the respondents. Indeed, the very order in which the options are listed and the range of options offered are likely to affect the response. Moreover, a range of optional answers from very negative to very positive may reduce a complex reaction to a superficial response. There is also a tendency to 'heap' on the middle point of the scale.

The designer must guard against the disguised multiple question and the excessively hypothetical question. Examples of these are:

"Do you normally work on your farm, but also seek employment on other farms?"

"Do you consider that your sons will be able to make a living from farming when they are older?"

The first example, in effect, asks two questions. Suppose the answer is 'no'; what interpretation can be put on the answer?

The second involves the respondent visualizing a situation in which various possibilities of change or lack of change will be adopted in the visualization process, and as these are not revealed in the recorded response, the answer may be meaningless. Both these questions and similar variants have been used in recent years.

The best attitudinal enquiries lead the respondent from a statement of fact regarding his behavior, to his reasons for acting as he did, to the circumstances that he thinks will lead him to change his behavior.

The Verbatim Questionnaire

This type removes all enumerator discretion. The questions are printed in the precise wording to be used during the interview; the options and branching processes that may occur during the interview are identified and specific

---

11. It is not only technical terms that give difficulty. Words such as "since" and "ever" are often mistranslated or misinterpreted.

FIGURE 7
AN EXAMPLE OF A PRE–CODED FORM

NATALIDADE

Se nos dois últimos anos alguma mulher desta casa ficou grávida, tome os seguintes dados:

1. Idade da mulher ao engravidar pela última vez nesse periodo.	1 2 3 4
2. Idade da mulher ao engravidar pela primeira vez.	5 6 7 8
3. Número de gestaçocs anteriores.	9 10 11 12
4. Número de abortos.	13 14
5. Número de natimortos.	15 16
6. Número de crianças falecidas antes de 1 anos.	17 18
7. Número de crianças falecidas de 1 a 5 anos.	19 20
8. Número de crianças falecidas com mais de 5 anos.	21 22
9. Problemas na ultima gestaçào (78 ou 79) (1. hemorragia; 2. infecçào; 3.eclampsia; 4. aborto; 5 outro: especifique).	23 24 25 26
10. Fase da gestaçào cm que ocorreu o problema (1. até o 3. º més; 2. no curso do 4. mès; 3. no curso do 5. més; 4. no curso do 6. º més; 5. no curso do 7. º més; 6. no curso do 8º més 7. no curso do 9. º més)	27 28 29 30
11. Em que fasa da gestaçao iniciou o pré-natal, na ultima gestaçao (78 ou 79) (1. Antes do 3. º més; 2. no curso do 4. º més; 3. no curso do 5. º més; 4. no curso do 6. º més; 5. no curso do 7. º més; 6. no curso do 8. º més; 7. no curso do 9. º més).	31 32
12. Número de filhos no último parto, dentro dos últimos 2 anos.	33 34
13. Nascidos (1. vivos; 2. mortos; 3. 1 vivo e I morto).	35 36
14. Locai do parto (1. residéncia; 2. maternidade localizada no município; 3. maternidade localizada no município vizinho; 4. maternidade localizada em município nào vizinho; 3. a caminho da maternidade; 6. fora da residéncia).	37 38
15. Parto assistido por: (1. médico; 2. pessoal paramédico; 3. agentes leigos de saúde; 4. outras pessoas).	39 40

instructions included in the questionnaire; in extreme cases, even the courtesy opening and closing of the interview are indicated.

This type of questionnaire is best suited for a detailed interview into a specific topic where the range of behavior is both limited and identifiable in advance. Its widest use has been in surveys of demographic characteristics, particularly fertility, for which it seems well-suited. It is not usually appropriate in a general farm survey, where enumerator reaction to the widely varying farm characteristics and farm patterns is almost essential, if the interview is to appear to be an intelligent one, as viewed by the respondent.

The decision to adopt this type of questionnaire carries the following implications:

(a) The questionnaire must be printed in the language of interview;
(b) The possible courses of the interview must be fully foreseeable; and
(c) The length of the questionnaire (if not the actual interview) is a major design problem.

The need for the questionnaire to be printed in the language of the interview is obvious; the perceived need for removing enumerator discretion cannot otherwise be achieved. Unfortunately, in certain countries, this may entail a translation of the questionnaire from the language used for drafting into, not one, but several languages. This is a tedious business and must be carefully done and tested. Indifferent translation produces the same result as enumerators exercising their discretion in phrasing the questions. Moreover, if technical words are used and not understood by the respondent, the enumerator is forced to clarify the meaning: this again defeats the purpose of structuring the interview to avoid enumerator bias. A common safeguard is to translate the questions into the required language and then to have this version independently translated back into the original in order to reveal the deficiencies and ambiguities.

All possible courses that the question and answer process may take must be covered by written instructions and alternative questions. In so doing, the questionnaire is likely to become formidably long, even if, in any one interview, the number of questions put are limited. Figure 8 shows an example of an acceptable extract from a structured questionnaire. Bad examples are, however, more common than good.

Enumerator Training Manual

A training manual must be prepared to accompany each questionnaire. This will be the basic reference document during the enumerator training course. It should contain the following:

(a) explanatory details about the survey,
(b) instructions for identifying the respondents,
(c) general guidelines on conducting an interview;

(d) precise definitions of each term used in the questionnaire, e.g., household, farm, field, income;

(e) explanation of each question with examples of how answers are to be recorded;

(f) coding instructions (if these do not appear on the questionnaire); and

(g) other operational and time-scheduling instructions.

It is important to stress that by the time the training course is completed, items (b)—(e) must be so familiar to the enumerator that reference to the manual during an interview is avoided. The manual will, however, continue to be a valuable reference document for the enumerator, particularly with regard to items (f) and (g), and also when a difficulty arises regarding interpretation of the definitions in an unexpected situation.

FIGURE 8
AN EXAMPLE OF A STRUCTURED QUESTIONNAIRE

1. Did you do any work in the last 7 days?

   Yes...1

   No....2   SKIP TO 6

2. What was that work?

   _____

   _____

3. How many hours altogether did you work over the last 7 days?

   _____
   HOURS

4. Did you want more hours of work during those 7 days?

   Yes...1   SKIP TO 9

   No....2

5. Why did you not want more hours of work?   CIRCLE ONE ONLY

   Worked fully 1
   At shool 2   ALL
   Housekeeping 3   SKIP
   Too young/old 4   TO 9
   Physical defect 5
   Other (SPECIFY)...... 6

6. Do you have a business for which you did
   anything during the last 7 days?

   Yes...1   SKIP TO 9

   No....2

7. Did you look for work during the last 7 days?

   Yes ...1   SKIP TO 9

   No 2

8. Why were you not working during the last 7 days?   CIRCLE ONE ONLY

   Vacation.............. 1
   Illness 2
   Strike 3   ALL
   Lay-off 4
   Off-season 5
   Other (SPECIFY)...... 6

108

Part 9:
DATA HANDLING AND ANALYSIS

9.1   Introduction

As stated at the beginning, this Handbook is designed to serve different types of users, and therefore contains material at different levels. That is particularly true of this Part. Early sections discuss simple first stage steps, treated at a very elementary level, and may be useful for the essential early stage of exploratory analysis before deciding on the details of more formal analysis, whether computerized or otherwise. Later sections touch on problems that are often apparent only to those with an advanced statistical background. The material has been selected with two ends in view:

(a) to provide examples of simple data handling techniques, relevant to the major theme of this Handbook of turning data into information for management; and

(b) to indicate some of the major sources of difficulty in analysis, thus providing a set of warning lights so that false conclusions will not be drawn and professional statistical advice will be sought when it is needed.

Much can, in fact, be achieved by simple techniques of analysis—the preparation of graphs, the calcuation of simple rates and averages, the ranking of items, and the construction of simple tables. These are particularly relevant to quick, timely reporting for monitoring. Exploratory analysis of this kind is also a required first step before more complex techniques are employed. These more complex procedures, primarily multivariate analysis, are almost certain to be required when casual analysis is involved, especially in the later stages of evaluation. Professional statistical advice will be required to decide what should be done and—often just as important—what should not be done. The availability of package programs is now so general that data are frequently subjected to inappropriate manipulations, results being read off computer printouts without adequate consideration of whether the methods of analysis are appropriate, given the character of the data and the methods by which they have been collected.

Simple techniques are important for first assessments of the quality of the data, and for establishing orders of magnitude of the main features under observation. These can also be used to give a first idea of the sensitivity of the results to variations from the norm.

Different types and scales of enquiry generate different requirements for data editing and coding; the following sections should be interpreted with that in mind. Experience shows that the data handling and presentation stages in the generation of information are difficult to complete in practice. There are numerous surveys and enquiries where the data collected have never been properly analysed and used. Some of the reasons for this are

(a) Insufficient attention is given to these later stages when the enquiry is being planned and the questionnaire designed; the main lines of analysis and presentation must be thought through before the data are collected;

(b) Failures of the kind described in (a) lead to the collection of much unnecessary data, which complicate the data processing and delays work on the data of primary concern;

(c) Expectations of speedy analysis arising from developments in electronic data processing are usually overoptimistic;

(d) The tedious nature of the work of editing and coding data dampens commitment and enthusiasm; leading to a tendency to do more field work before bringing the current study to a conclusion;

(e) The task of constructing and then validating large data files, particularly those relating to longitudinal surveys, is usually grossly underestimated.

9.2 Data Preparation

An editing stage is necessary for all but the smallest investigations. Data are edited for completeness and consistency; omissions and inconsistencies discovered are re-enumerated in the field or corrected by using an agreed imputation process.

If the proposals made in Part 8 have been followed, office coding will have been reduced to a minimum through the use of precoded questions. The resulting coded data are either transcribed directly on to summary tables, or transferred to cards, discs or tapes for further analysis.

Unless office management is strict, this work can degenerate into chaos. The questionnaires must be booked in, grouped into appropriate enumerator, area, or strata bundles, and their movement between different stages and clerks recorded. Detailed records must be kept of any questionnaires returned to the field or temporarily abstracted for other purposes. Any dislocation of the normal process by which batches of questionnaires pass through manual edit, correction, and punching should be avoided.

Frequent inspection of daily output figures are helpful in monitoring staff performance. Manually maintained running totals of some of the key variables can be used to provide averages, rates, orders of magnitude, and first estimates of the main characteristics under enquiry. If there are previous or preliminary figures with which these first estimates can be compared, discrepancies may provide an early warning that something has gone wrong.

It is important that the person in charge of the enquiry should examine and edit a sample of the field documents as they arrive. This provides a check on earlier assessments of the quality of the field work, and ensures that the editorial instructions to detect inconsistencies are sound. In fact, unless a pre-test is conducted, it is often not possible to complete the instructions for editing and coding until some field documents have been scrutinized.

Not all errors require the return of questionnaires to the field for comment, revisit, and amendment. Some are obviously slips and can be corrected in the office. Others, such as incorrect area measurements cannot be corrected without further field enquiry, which must be undertaken promptly. Field checks are also required for some information which appears dubious but is not necessarily wrong.

'Faked' data is not uncommon, for most survey teams contain a few dishonest enumerators who manage to survive the training course and its selection filter. This is one reason why it is desirable to batch questionnaires by enumerator since a skilled editor can often detect patterns in the responses, e.g., a suspicious regularity in family structures, or the absence of occasional zeroes, which indicate that faking may have occurred.

Editing, particularly in the early stages, should also show whether there has been a misunderstanding of procedure that has escaped the supervision in the field. For example, suppose it was decided that in the case of an absentee husband, the wife was to be shown as the head of the household: an absence of questionnaires showing wives as heads of households in any area may suggest that this rule has not been followed there.

Editing and coding stages are sometimes combined. There are advantages and disadvantages in this arrangement; on balance, unless the questions are almost all precoded, it is desirable to separate the jobs. Quality control—a regular check of a sample of forms after editing and coding—is needed to ensure that the work is being done properly. This check needs to be maintained throughout the process, and may have to be stepped up in the later stages as earlier enthusiasm and commitment ebb away.

9.3   Data Processing

The equipment and their associated programs for data processing are rapidly changing. In particular, development of microcomputers seems likely to alter the situation considerably. This development may prevent the 'distancing' of the surveyor from the data that occurs with main-frame installations. With a microcomputer the surveyor is more in control of the processing; in control of access to, as well as design of, analysis. The design itself can be more flexible. The surveyor can more easily control the structure and validation of the data files and so avoid some of the numerous validation runs that are characteristic of present electronic data processing. On the other hand microcomputers are as yet limited in their ability to handle large files, and software development lags behind that of the hardware.

In the current situation of rapid change, detailed recommendations would be out of place. Two comments should, however, be made. First, the computer is not the *only*, or necessarily, the best way in which data can be processed. Much can be done with data summarized by appropriate groups and analyzed with the aid of cheap, programmable, hand-held calculators. When timeliness is essential—and this is often the case in monitoring—some simple tabulations from the data, or a sample of it, will often meet the requirements effectively.

If the regular data for monitoring purposes can be processed through available computer facilities, it will be well worth the initial investment to get the system working; this is one aspect of the general use of the computer in management. However, operations at distant sites with difficult climates may face problems. Further, the temptation must be resisted to expand reports excessively simply because the facilities are there. Experience varies from project to project, and even similar projects in the same country show considerable variations in performance which are difficult to account for.

Secondly, the time allowed for processing survey type data is usually too short, particularly when computer processing is involved. In a recent survey of data processing of the World Fertility Survey, Rattenbury writes that,

Realistic estimates, to allow for shortfalls in performance and for unpredictable problems, are obtained by multiplying the normal estimates by a realism or R-factor. From experience in many environments this factor should often be three or more for data processing projects. Actual durations of WFS country data processing have been on average about three times greater than planned.[12]

The World Fertility Survey was usually given some priority in the country concerned, and substantial use was made of the additional data processing skills of the international survey staff.

9.4   The Appropriate Scale of Analysis

In many cases analysis will not be a problem, since the basic data are limited and simple, and the techniques required straightforward. For many aspects of monitoring, the data processing requirements will be laid down in a reporting sequence repeated regularly throughout the year. The analysis of larger and more complicated sets of data resulting from case studies and sample surveys, however, require decisions involving the crucial trade-off between speed and depth.

---

12. J. Rattenbury, "Survey Data Processing— Expectations and Reality," *World Fertility Survey Conference* (London: July 1980).

Reports of past experience emphasize that timeliness has generally been given insufficient weight; this Handbook has accordingly tried to redress this by emphasizing the value of quick, simple summaries. More sophisticated analyses will, however, be required when causal links are being investigated.

Statisticians have debated the relationship of levels of data accuracy and levels of analytical detail and sophistication. Some have argued that sophisticated methods should not be applied to bad data; others have argued that these are especially required when the data are bad. A 'middle of the road' position is

(a) No general rule can be given: each case must be considered on its own merits—at all costs, avoid the "tabulate everything against everything" attitude;

(b) It is sometimes worth trying sophisticated methods to see what can be 'squeezed' out of the data. The results must, however, be interpreted with great care. For example, the signs and magnitudes of the estimates of the coefficients for some of the variables in a multiple regression relationship may be different from expectation. If these are ignored and attention concentrated on those which confirm expectation, the analysis is being used merely to support preconceived ideas and does not provide an independent validation;

(c) The extent to which the results are sensitive to the likely size of inaccuracies should be investigated by simulating alternative sets of data;

(d) The necessity for effecting data transformations should be examined;

(e) Data dredging is dangerous, and will almost certainly generate spuriously statistically significant results. On average, one in twenty tests of nonsignificant relations (using 95 per cent confidence limits) will appear significant; and

(f) Statistical significance does not necessarily imply substantive significance.

Some effects of inaccuracies in the data are set out in Section 9.9.

9.5 Presentation and Display

The manner in which data is presented depends upon the aspects that need highlighting. A common fault is the failure to simplify: usually, far too many digits are given. Partly, this is because, say, a figure of 71654 under a heading of 'Tons' gives a reassurance of accuracy not provided by an entry of 72 under a heading of 'Thousand Tons': it is also due partly to an inheritance from bookkeeping, where an exact balance must be shown.

Detailed figures must, of course, be available and supplied to those who need to make further computations. But management first wants to see the main outlines of the situation; tables and graphs must be designed for this purpose. Some of the possibilities are illustrated using the data in Tables 3 and 4, taken from the detailed report of a survey.

Table 3: Maize and Groundnut Yields and Gross Production
for the 1971 Harvest Periods

Area	Maize		Groundnuts	
	Average Yield Per Acre	Gross Production	Average Yield Per Acre	Gross Production
	(lbs.)	(tons)	(lbs.)	(tons)
1	1,331	12,772	476	2,215
2	1,230	9,284	360	2,227
3	1,110	14,135	520	4,816
4	1,290	13,041	580	5,864
5	1,090	8,073	480	3,555
ALL AREAS	1,209	57,305	494	18,676

Table 4: Maize and Groundnut Yields and Gross Production
for the 1972 Harvest Periods

Area	Maize		Groundnuts	
	Average Yield Per Acre	Gross Production	Average Yield Per Acre	Gross Production
	(lbs.)	(tons)	(lbs.)	(tons)
1	1,631	13,989	297	1,640
2	1,628	12,288	363	1,887
3	1,845	21,005	514	4,958
4	2,038	19,027	452	3,515
5	1,020	7,414	367	2,823
ALL AREAS	1,626	73,723	413	14,824

The presentation could be varied to highlight different aspects. For example, if a comparison for maize is required, an obvious first stage would be to reduce the number of digits and produce Table 5.

Table 5: Maize: Average Yields per Acre and
Gross Production, 1971/72

Area	Yield (hundred lbs. per Acre)		Production (thousand tons)	
	1971	1972	1971	1972
1	13	16	13	14
2	12	16	9	12
3	11	18	14	21
4	13	20	13	19
5	11	10	8	7
ALL AREAS	12	16	57	74

The changes can now be seen at a glance, e.g., that the variations between areas increased markedly from 1971 to 1972.

Table 5 could be extended to include a measure of change. Change can be measured in absolute or relative terms, e.g., the absolute change in maize yield in Area 1 is 300 lbs. (1631 - 1331), and the relative change is 23% (100 x 300/1331). Both measures are useful, but the relative change is often the more relevant for comparative purposes. Table 6 presents results of this kind.

Table 6: Maize and Groundnuts: Percentage Changes in Average
Yield per Acre and in Gross Production, 1971-72

Area	PERCENTAGE CHANGE			
	Maize		Groundnuts	
	Yield	Production	Yield	Production
1	23	10	−38	−26
2	32	32	0	−15
3	66	49	−1	+3
4	58	46	−22	−40
5	−6	−8	−24	−21
ALL AREAS	35	29	−16	−21

A succession of tables in a major report often need to present areas in a regular order to avoid confusion. However, the relative performances of a number of areas may be brought out by ordering them according to the extent to which they deviate from the average level of performance. Thus a 10 area project might produce the following results:

Area	7	4	10	1	8	9	5	2	3	6
Performance (%)	50	55	55	60	65	70	75	80	95	105
Deviation from average (71%)	−21	−16	−16	−11	−6	−1	+4	+9	+24	+34

Some common characteristics of areas close together in the ranking may suggest management action. For example, if areas 3 and 6 are following some practice not adopted elsewhere, perhaps it should be introduced in other areas.

It should be noted that overperformance—getting results higher than targets—requires reaction, just as much as underperformance. If credit operations in one area exceed those planned for, extra supplies of fertilizers, seeds, etc., will be needed there.

Absolute figures of area targets and performances can be plotted using the two axes to represent target and performance respectively. The data in Table 7 are shown in this graphical form in Figure 9:

Table 7: Target and Performance by Area

Area	Target	Performance	Area	Target	Performance
1	24	35	6	51	59
2	102	112	7	38	25
3	65	68	8	140	82
4	132	105	9	96	62
5	100	90	10	114	95

Any area where performance equalled target would be on the diagonal: points below the diagonal show areas which have not reached their targets, whilst those above show those where more has been done than expected. The length of the vertical deviation of the point from the diagonal represents the extent of the shortfall (or of overperformance). In this example the areas with the larger targets are underperforming, and their shortfalls are substantial.

It is not always necessary to react to short-term fluctuations within individual reporting units. For these cases cumulative performance graphs bring out the situation in a useful form. Three profiles are shown in Figure 10. They are especially simple in the sense that they assume targets increasing linearly over time (that is, increasing by the same absolute amount for equal time periods). Profile (i) shows an operation that started off reasonably well, but then deteriorated. It also shows how warning lines may be inserted, and action initiated when operations fall below some control level. More complicated "cusum" (cumulative sum) techniques are available as part of statistical quality control. Profile (ii) indicates an operation that started badly, but then got "on course." As the opportunities lost at the beginning could not be retrieved, a new set of targets was fixed. Profile (iii) shows an oscillating performance. It may be noted that the slope of the between-period line indicates whether the

FIGURE 9
TARGET AND PERFORMANCE BY AREA

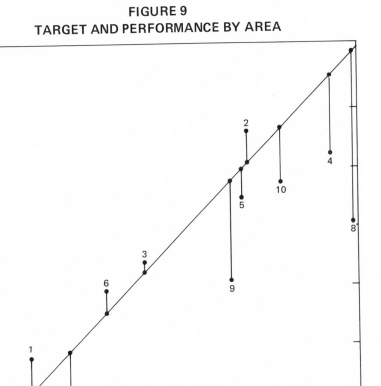

117

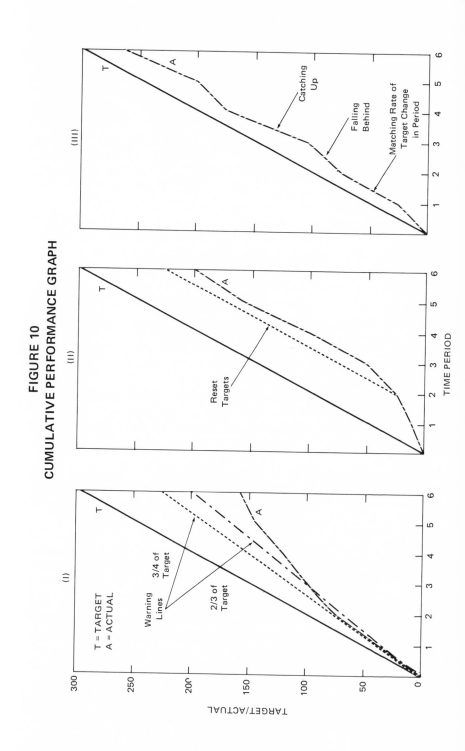

FIGURE 10
CUMULATIVE PERFORMANCE GRAPH

118

operation is falling further behind (slope of performance line less than that of the target line), holding its own (equal slopes - or parallel lines), or catching up (slope performance line greater than that of the target line).

The three examples start with the same target. Graphs of this kind can be produced in percentage form (target at end of final period = 100%) and will then be comparable between reporting units even if their targets differ in absolute size.

In many cases, interest is focused on the rate of change, not on the absolute amount. If a time series is graphed using a logarithmic scale on the vertical axis, then a straight line represents equal relative increases per unit of time—a constant rate of change. Such a graph is called a semilogarithmic graph (in casual reference the 'semi' is often omitted). If semilogarithmic graph paper is not available ordinary graph paper can be used by first converting the data into logarithms and plotting the log values rather than the absolutes. Negative numbers cannot be accommodated using this approach as logarithms of these do not exist.

Semilogarithmic graphs are also useful when the scale for the variable plotted up the vertical axis covers a very wide range, e.g., from 100 to 100,000. This situation often occurs when comparisons are being made between regions of very different sizes.

When examining the relationship between two variables, one representing a cause and the other an effect, a scatter diagram can be used as a visual aid. Thus if yields are plotted against amounts of fertiliser used, the resulting diagram may enable any relationship or lack of it to be easily seen. This is an useful first step before proceeding to a more formal relationship analysis using regression techniques (discussed below). In particular, a visual examination of a scatter diagram may prevent an inappropriate regression analysis being performed.

9.6 Description and Measurement of Inequality

Inequality and the extent to which it is changing are often of concern in agriculture and rural development projects.

The simplest measure is the proportion of the benefit being distributed in the possession of the poorer part of the population. Thus, it may be said, that 60% of the farmers own 20% of the cultivable area. This is obviously a crude measure, since changes of the distribution towards either greater or lesser equality can take place without the figure at the cut-off point chosen being affected appreciably.

For comparative purposes, inequality may be summed up by the Lorenz diagram, and the Gini coefficient, associated with it. These, in effect, combine a series of percentage relationships of the kind mentioned above. The data are put into cumulative percentage form as in the following example, using data of the distribution of farms by area (Table 8). A Lorenz diagram is obtained by plotting these cumulative numbers, as shown in Figure 11. The extent of the inequality is indicated by the shaded area between the diagonal and the curve: the greater this area the greater the inequality.

Table 8: Distribution of Farm Size

Size Class (areal units)	Cumulative Percentage of Farms	Cumulative Percentage of Area
0 - 1.0	4.4	0.7
1.1 - 2.0	19.0	5.4
2.1 - 3.0	33.3	13.1
3.1 - 4.0	45.5	21.6
4.1 - 5.0	61.5	36.0
5.1 - 6.0	70.1	46.1
6.1 - 7.0	78.8	56.0
7.1 - 8.0	84.9	65.0
8.1 - 9.0	87.8	69.7
9.1 - 10.0	92.3	78.1
10.1 and above	100.0	100.0

The Gini coefficient is the proportion that this shaded area bears to the area of the whole triangle in which the curve lies. It can be calculated as follows:
(a) take the sum (Tl) of cross multiplications of the following kind:

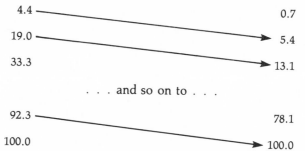

. . . and so on to . . .

(b) Take the sum of (T2) of cross multiplications in the opposite direction:

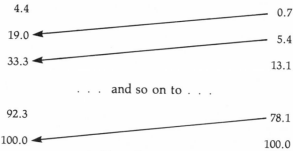

. . . and so on to . . .

(c) The Gini coefficient $= \dfrac{T1 - T2}{10,000}$

FIGURE II
LORENZ CURVE FOR DATA IN TABLE 8

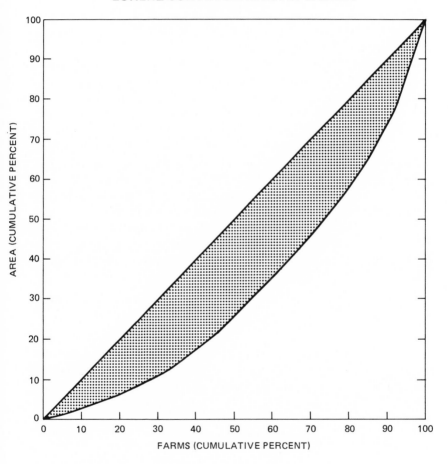

It must lie between 0 and 1, and in this case is 0.354.

It should be noted that the first column of cumulative percentages relates to the units (farms, persons, etc.) between which the 'benefit' (area, income) is distributed. The second column is the percentage of the 'benefit' owned by the related units in the first column. Entries in the first column are plotted along the horizontal ("x") axis in the Lorenz diagram; those in the second column along the "y" axis. The same scale is used for both axes.

If two distributions are being compared, the distribution with the larger Gini coefficient is the more unequal. But this rule has to be interpreted with care because two types of situations can arise.

(a) The Lorenz curve for distribution A lies entirely within that for distribution B. Then the conclusion that distribution B is more unequal than distribution A is unequivocal.

(b) The Gini coefficient of distribution B is greater than that of distribution A, but the Lorenz curves of the two distributions cross each other when plotted on the same diagram. This indicates that distribution B is more unequal than A over part of the range, but less unequal over another part. Conclusions about relative inequality will depend upon whether the whole distribution, or only part of it, is the focus of interest.

Gini coefficients provide a convenient first ranking of inequalities, but too much should not be made of small differences between them. They will be affected by response errors in the data, which are often likely to be large for variables used to investigate inequality; the way the data have been grouped will influence the values; and if sample data have been used, sampling fluctuations will be present. Thus, whilst Gini coefficients are useful indicators, conclusions about significant changes in relative inequalities must rest upon detailed examination of the original distributions.

9.7   2 x 2 Tables

One of the commonest ways of showing differences in performance is by a 2 x 2 table. A typical example is given in Table 9 (a). The letters in Table 9 (b) will be used for generalizing results.

Table 9: (a) Adoption of Practice by Project and Nonproject Farmers

	Adopt	Do not Adopt	Total
Project farmers	180	70	250
Non-project farmers	60	90	150
Total	240	160	400

	Yes (1)	No (0)	Total
In (1)	a	b	(a+b)
Out (0)	c	d	(c+d)
Total	(a+c)	(b+d)	$n=(a+b+c+d)$

The difference in the proportions 'adopting' for project and nonproject farmers gives an idea of project performance. The difference in proportions is:

From Table 9 (a)

$$p_1 - p_2 = \frac{180}{250} - \frac{60}{150}$$

$$= 0.32$$

From Table 9 (b)

$$p_1 = a/(a+b); \quad p_2 = c/(c+d)$$

$$p_1 - p_2 = \frac{a}{a+b} - \frac{c}{c+d} = \frac{ad - bc}{(a+b)(c+d)}$$

If several tables are being compared, comparisons of p_1's and p_2's can be made. Simple single figure measures have however been devised.

One single figure method is the relative odds or the Odds Ratio (OR)— the ratio between the odds that a project farmer has adopted to the similar odds for a nonproject farmer. For Table 9 the results are:

From Table 9 (a)

$$OR = \frac{180}{70} \div \frac{60}{90} = \frac{180 \times 90}{60 \times 70}$$

$$= 3.9$$

From Table 9 (b)

$$OR = \frac{a}{b} \div \frac{c}{d} = \frac{ad}{bc}$$

The bigger the OR, the greater the odds for the first group, relative to the second.

The values that the Odds Ratio can take range from 0 to infinity. Values 0 to 1 indicate a negative relationship—nonproject farmers are adopting more than project farmers. Values greater than 1 indicate a positive relationship.

For making comparisons across a number of tables a measure that ranges from -1 to $+1$ may be more convenient. One such is Yule's Q, which is calculated as follows:

$$Q = \frac{180 \times 90 - 60 \times 70}{180 \times 90 + 60 \times 70}$$

$$= 0.59$$

From Table 9 (b)

$$Q = \frac{ad - bc}{ad + bc}$$

Q was originally devised as a measure of association, and is appropriate here since a difference in the odds on adoption for project and nonproject farmers is equivalent to an association between "being in the project" and "adopting the practice."

Q takes values ranging from 0 (no association) to +1 in one direction (complete positive association) and -1 in the other (complete negative association). The larger the figure, the greater the strength of the association it summarizes. Q and OR are related in the following way:

$$Q = \frac{OR - 1}{OR + 1}$$

An OR of 3 is equivalent to a Q of 0.5.

Another measure sometimes used is the correlation coefficient r (which also takes values lying in the range -1 to $+1$). The usual formula for r can be put as follows (using the assignment of the numeric codes 0 and 1 in Table 9 (b)):

From Table 9 (a)

$$r = \frac{12000}{\sqrt{(250 \times 150 \times 240 \times 160)}}$$

$$= 0.316$$

From Table 9 (b)

$$r = \frac{ad - bc}{\sqrt{[(a+b)\ (c+d)\ (a+c)\ (b+d)]}}$$

The result for r will be of the same sign as Q (since the numerator is the same, and the positive root is always taken in the denominator of r) and r will always be closer to 0 than Q. It is often recommended (see 9.9), that the value of r^2 be considered, rather than that of r, since r^2 equals the proportion of the variation explained. The value of r^2 in this case is equal to only 0.1.

The correlation coefficient is, however, better suited to quantitative than to qualitative variables. Any pair of quantitative variables might be turned into two zero-one variables by splitting each at a point. For example, distributions of expenditures of farmers and of the areas they farm can provide the basis for a 2 x 2 table by forming groups labelled:

Expenditure	Farm
High	Large
Low	Small

Values of Q could then be calculated, but they are sensitive to the choice of the splitting points.

The quantity (ad − bc) occurs again in another often used measure. Suppose there is in reality no association between the two variables. In this case it would then be expected that the (a + b) farmers and the (c + d) farmers, and the total of farmers would all divide in the same proportion. In the cell in which a is found, the number would be expected to satisfy:

$$\frac{a}{a + b} = \frac{a + c}{n}$$

giving:

Expected value in the first cell under assumption of no association $= \dfrac{(a + b)(a + c)}{n}$

The number actually observed is a, and so the difference (D_{11}) between the observed and 'expected' values is:

From Table 9 (a)

$$D_{11} = \frac{(180 \times 90) - (60 \times 70)}{400}$$

$$= 30.$$

From Table 9 (b)

$$D_{11} = a - \frac{(a + b)(a + c)}{n}$$

$$= \frac{ad - bc}{n}$$

Keeping the marginal totals constant, it will be seen that $D_{11} = D_{22} = -D_{12} = -D_{21}$.

Clearly, the larger (ad − bc)/n, the bigger the difference between the observed value and that expected on the assumption of no association, the stronger is the evidence for the alternative view that there *is* a relationship between the variables.

Using this feature, a commonly used test for assessing sampling fluctuations, called 'chi-squared', is calculated as follows:

From Table 9 (a)

$$X^2 = \frac{400(12000)^2}{250 \times 150 \times 240 \times 160}$$

$$= 400(.316)^2$$

From Table 9 (b)

$$X^2 = \frac{n(ad - bc)^2}{(a+b)(a+c)(b+d)(c+d)}$$

or substituting r from above,

$$= nr^2$$

If there were no association between the variables, samples giving a value of X^2 greater than 1 would occur, on average, approximately once in three samples, and values greater than 3.8 would occur only once in twenty times. One reason for calculating this particular quantity is that it is a convenient basis for combining results from a number of similar sample tables. If the association in all, say k, tables is in the same direction (that is if (ad − bc) is the same sign in all k tables) then the sum of X^2's from the tables can be added together and tested at the desired level against the value of X^2 with k degrees of freedom (details in most sets of statistical tables).

Single figure measures provide only a first stage of analysis, and the reduction of the data into 2 x 2 tables may be an oversimplification. It should also be noted that the description of these measures are introductory only. The effects of sampling design features, e.g., clustering, have not been dealt with. But an initial analysis of this kind can be done quickly by hand and be valuable in indicating whether there is a possible relationship that will warrant a more detailed analysis under professional guidance.

9.8 Standardization

Standardization is a procedure which compensates for differences in structure between groups or areas being compared. Suppose that the readiness and ability of farmers to take up credit is related to the size of their farms, those farming larger areas being on average more likely to take up credit. Differences in the overall percentage absorption of credit in project areas could be due to differences in administrative effectiveness or to differences in the distributions of farms by size (or, of course, to other factors which are ignored here). Table 10 gives a numerical example:

Table 10: Differences in Area Performance due to Structural Differences

Size of farm	Area A			Area B		
	Proportion (1)	% take-up (2)	Product (1)x(2)	Proportion (3)	% take-up (4)	Product (3)x(4)
Small	0.50	18	9	0.33	18	6
Medium	0.25	36	9	0.33	36	12
Large	0.25	72	18	0.33	72	24
Overall % take-up			36			42

Area B's overall rate of 42% looks better than Area A's rate of 36%, but the difference is due to the higher proportion of smaller farms in Area A.

Standardization deals with this problem by comparing the overall results with that which would have been obtained if the detailed rates were matched with a standard structure, which is usually the structure for the whole region under study. Thus, supposing the proportions of farms by size in the total project region were 0.5, 0.3, and 0.2, standardized figures would be calculated as follows (first three areas only shown):

Table 11: Calculation of Standardized Rates

Size of farm	Standard Proportion	Area A		Area B		Area C	
		% take-up	Product	% take-up	Product	% take-up	Product
Small	0.5	18	9.0	18	9.0	20	20.0
Medium	0.3	36	10.8	36	10.8	35	10.5
Large	0.2	72	14.4	72	14.4	75	15.0
Overall % take-up (standardized)			34.2		34.2		35.5

The overall standardized rates for Areas A and B are equal, as they should be. These standardized rates can also be used to provide measures of relative performance by comparing them with overall project area percentage take-up rate. Suppose this is 30%. Then a suitable index of performance is:

$$\text{Area A} \quad 34.2/30 \times 100 = 114$$

$$\text{Area B} \quad 34.2/30 \times 100 = 114$$

$$\text{Area C} \quad 35.5/30 \times 100 = 118$$

Standardization can be extended to cover more than one structural feature, if desired. The procedure was developed for the comparison of mortality rates and is formally similar to that used in the construction of index numbers.

9.9 Cautionary Comments About Correlation and Regression Analysis

Attempts to trace causal relationships in a project will often use correlation and regression methods. Some of the dangers they involve are briefly outlined in this Section, so that due caution will be exercised in drawing conclusions from them.

There is a time-honoured methodological maxim: "correlation does not prove causation." Correlation of an effect with an identified treatment is only useful evidence if the results of other possible causes have been partialled out. Thus, for example, although production may have increased, the increase may be due to better market prices, rather than to the additional fertilizer. Conversely, a project failure may be due to an unforeseen change in general price structure, not to any internal fault.

The correlation of time-series often leads to mistaken inferences. A high figure may be due merely to the separate effects of time on the two series, not to any causal relationship between them. It is therefore usually worthwhile to see if the relationship is maintained when the two series of first differences (that is, the period-to-period changes) derived from the original data are also correlated. If the value of this second r is also high, confidence that there is a real relationship between the two series is increased.

A study of the residuals (the differences between the original observations and their estimates obtained from the model being considered) is necessary to decide if the model is applicable. Patterns shown by them may indicate how the model can be improved: for example, if the original observations are spread over time and the residuals show a trend over time, then adding a time term to the model may improve it. Further, correlation coefficients measure *linear* correlation. A close nonlinear correlation may be hidden in a misleadingly low value of the coefficient. The data and residuals must always be examined to see that any assumption of linearity is appropriate.

A second reason for looking at the residuals is to see that any assumptions about the error structure of the model are met. For example, the classical linear regression model requires that the variability of the dependent variable is constant across the range of values of the independent variables. If the residuals indicate this is not so, then some transformation of the independent variables may be necessary. Tests of significance and confidence intervals usually require that the residuals in the model are distributed normally—at least approximately. So a study of the residuals can prevent improper inferences.

Another common error is referred to as the 'ecological fallacy', and arises when correlations between variables measured over aggregates of persons do not appear when the correlation is calculated over individuals. Thus, there may be correlation at an aggregated level (say, by district) between educational attainment and business activity; but when the data is looked at on the individual level there may be little correlation between education, and commercial or industrial initiative, and ability. Indeed, a general problem with correlation coefficients is that their size can be increased by aggregating upwards the data used to calculate them.

In any case, the value of r is often not the most relevant figure. A value of 0.5 for the correlation between treatment and impact indicators may seem high; but it should be regarded as showing that the treatment explains 25 percent, $(0.5)^2$, of the variation in the impact indicator. If $r = 0.2$, still some way from zero, the treatment is explaining only 4% of the variation. Too much attention may also be given to a finding of 'statistical significance'. This is often merely a flag indicating that the effect under investigation is not likely to have arisen just by chance. This flag does not necessarily give any guide to the substantive or policy significance of the relationship: it may just prevent the waste of time that would occur in further examination of a chance relationship. Many correlation and chi-squared results found to be statistically significant are of little interest or importance for policy purposes.

Perhaps the most serious problem in using regression models in the context of this Handbook is the effect of measurement errors. Note that these errors are *not* sampling errors and that they can affect the *independent* variables. They are the difference between the true value for any respondent and the figure that is actually recorded and used in the analysis. The relationship involved is that the value included in the analysis for the i'th individual, x_i, does not necessarily equal the true value of X_i, the error being equal to e_i. Thus:

$$x_i = X_i + e_i.$$

The e_i's can be positive or negative, and can of course be equal to zero. Errors can be introduced during editing or coding, but they arise more commonly in the collection of data in the field. Much of the discussion in other parts of this Handbook is directed at minimizing these errors.

In a simple one-independent variable regression model, the 'true' value of its parameter b, is related to the calculated value $\hat{b}$, by the equations:

$$b = \frac{\hat{b}}{R} = \hat{b}\frac{1}{R}$$

where R is the 'reliability' of the measurement of the independent variable. This result is sometimes referred to as the 'attenuation' effect.

$$R = \frac{\text{variance of X}}{\text{variance of X} + \text{variance of e}}$$

The variance of e may be expressed as a proportion of the variance of X:

$$\text{var e} = k. \text{var X.}$$

Then $R = 1/(1+k)$, and $1/R = 1 + k$. Thus if $k = 0.1$, the calculated value, $\hat{b}$, should be multiplied by 1.1 to correct for the attenuation effect. The value of k may be as high as 0.2 or 0.3, particularly when attitudinal and perception questions or long recall periods are involved.

The situation is more complicated with multiple regression. As an example, Cochran calculated the following table:[13]

---

13. W.G. Cochran, "Errors of Measurement in Statistics," *Technometrics*, 10(4), 1968, p. 657.

Table 12: Values of $\hat{b}_1$, $\hat{b}_2$ When $b_1 = 2$, $b_2 = 1$

(R_1 and R_2 are the reliabilities of the two independent variables and r is their correlation).

R_2		$r = +.3$ $R_1 = 0.6$	0.8	1.0	$r = -.3$ $R_1 = 0.6$	0.8	1.0
0.6	$\hat{b}_1$	1.25	1.68	2.13	1.10	1.48	1.87
	$\hat{b}_2$	0.74	0.66	0.58	0.44	0.51	0.58
0.8	$\hat{b}_1$	1.20	1.63	2.06	1.13	1.52	1.94
	$\hat{b}_2$	0.99	0.89	0.78	0.59	0.69	0.78
1.0	$\hat{b}_1$	1.15	1.57	2.00	1.15	1.57	2.00
	$\hat{b}_2$	1.25	1.13	1.00	0.75	0.87	1.00

Even if the second variable is measured *without* error, it will be seen from the bottom row that $\hat{b}_2$ can still over *or* under estimate b_2.

Measurement errors add considerably to the problems of using ratio variables in regression models.[14] A recent review of this issue produced the following results:[15]

Table 13: Effect of Random Measurement Error on the Correlation Between a Ratio (Y/Z) and Its Common Component (Z) Where the two are Linearly Related

Error Variance in Z as % of Total Variance	Error Variance in Y None Correlation r	Slope b	Same Proportion as in Z Correlation r	Slope b
0.00	1.000	1.000	1.000	1.000
0.05	0.899	0.901	0.812	0.901
0.10	0.797	0.798	0.660	0.798
0.20	0.586	0.588	0.420	0.588
0.30	0.368	0.372	0.232	0.372
0.40	0.139	0.143	0.075	0.143
0.50	-0.101	-0.109	-0.065	-0.109
0.60	-0.346	-0.433	-0.203	-0.433
0.70	-0.351	-2.315	-0.316	-2.315

14. A ratio variable arises, for example, if the ratio of food expenditure (labelled Y, say) to total expenditure (Z) is regressed on total expenditure (Z)—that is, Y/Z is regressed on Z.

15. S.B. Long, "The Continuing Debate over the Use of Ratio Variables: Facts and Fiction", *Sociological Methodology*, 1980, p. 55.

When the measurement error is substantial it will be seen that both b and r (the estimate of r) have the wrong sign.

Unfortunately, it is difficult to estimate response errors in individual surveys without complicated schemes involving interpenetrating samples (see Part 7), a special section in the pilot survey, or a post-enumeration survey. For enumerators dealing with similar groups of respondents, the variability within enumerators' results for a variable may be compared with the variability between enumerators' average results. Details are complex and a discussion will not be attempted here.

In conclusion, it should be emphasized that correlation and regression methods remain essential. However, the accessibility of computers has led to a proliferation of superficial analyses—improperly screened data fed into package programs followed by mechanical interpretations of the results. The way in which even experts can disagree about the appropriate methods and the interpretations of results shows the scale of the problems; and the warnings in this section may help to prevent some mistakes.

PART 10:
DATA PRESENTATION

10.1   From Data to Information

The data collector and processor sometimes forget that the users require information, briefly and clearly presented, rather than large volumes of indigestible figures. Project management and planners reviewing project performance need decision-oriented information from monitoring and evaluation systems; it is the task of those responsible for these systems to turn data into such information.

Failure to effect this metamorphosis can range from complete break down to just missing the final step, as illustrated by the following examples:

(a) data remains on the questionnaires, unanalyzed and valueless;

(b) magnetic tapes containing large data files are prepared but they remain unusable due to lack of proper validation procedures or documentation;

(c) tabular printouts, large in volume, long in detail, lie in files gathering dust in the data library;

(d) reports contain adequately presented summary tables derived from a baseline survey, but are available to the user only at the end of the project; and

(e) reports are full of tests of significance, analyses of variance, correlation matrices, etc., but do not set out any conclusions or suggest a range of options for action.

Converting data into information for presentation to the user is facilitated if the following guidelines are observed:

(a) The appropriate levels of detail and disaggregation will vary according to the different levels of user;

(b) The definitions of variables and tabular headings must be clear to the user who will not always have either a numeric background or technical knowledge of the topics discussed;

(c) The depth of statistical analysis must be geared to the level of user;

(d) The tabular layout, including the use of averages, dispersion indices,

132

ratios, etc., should be simple and clear—a set of simple two-way tables may be better than a complicated four-way cross-classification with each cell containing entries for absolute numbers, percentages, means, etc.;

(e) Text accompanying tables should summarize the main highlights revealed by the tables, indicating the conclusions that may be drawn; and

(f) Graphical and other diagrams will be particularly useful in focusing the user's interest and aiding his understanding.

10.2 Presentation of Information

The following essential questions must be borne in mind:

(a) to whom is the material addressed?
(b) at what time and with what frequency is it required? and
(c) in what form should it be given?

The recipients of monitoring and evaluation information include

(a) the field-level official of an agency contributing to the implementation of a component of the project;
(b) the project manager;
(c) the interagency coordinating committee;
(d) the sectoral planner;
(e) senior officials at the national ministerial level; and
(f) financial and donor agencies.

In general, as one moves along this scale, the information presented needs to be briefer, less demanding in the technical knowledge required by the recipient, and more oriented to presenting conclusions and options for decisions. This is not to say that the sectoral planner, for example, may not then call for the original data files in order to carry out further detailed analyses. But, the further the user is removed from a specific project, the more likely it is that many similar reports of a variety of projects pass across his desk; the initial perusal to identify the ones requiring further study is assisted if they are succinct.

Timeliness at an agreed frequency is the key to successful information transmittal. The most informative report is useless if the decision to which it should have contributed has, of necessity, already been made. The frequently encountered lack of reliance by the project implementers on their monitoring or evaluation units is usually due to a lack of confidence, based on experience, that the information required will be available when it is needed. If the emphasis in this Handbook on simple data collection procedures using appropriate sources and methods is followed, timeliness will be more likely.

The availability of information at the right time is also facilitated if the information needs and their timing are identified at an early stage. If a longitudinal study is launched using large samples, it is clearly impossible to report much in the way of results for a considerable time. Perhaps a different type of survey was required. The results of an objective measurement of crop yields cannot, by definition, be made available in advance of harvesting. If crucial decisions must be made before the harvest, and these require estimates of likely production, it is a crop forecast survey that is required as a supplement to, or substitute for, the later, more accurate effort.

Frequency of information transmittal is a matter for agreement between user and analyst, although their agreement will be worth little if the data collection and analysis mechanism cannot be designed or adjusted accordingly. Simple information on routine implementation progress may be required weekly or monthly. Moving averages of input data may be reported monthly or quarterly. Seasonal or cyclical data surveys should be organized to facilitate reporting within a fixed period of the events being recorded, as agreed in advance to meet user needs. Case studies and ad hoc problem-oriented surveys will not fit into a regular reporting schedule, but must be operated against deadlines for completion.

Quarterly, half-yearly, or annual reporting will occasionally match the frequency required to meet main user needs. More often, such fixed reporting cycles are inappropriate for this purpose. Flexibility must be the keynote. Approximate estimates, handwritten if necessary, may be invaluable if available on schedule.

Above all, the detection of an unexpected phenomenon, whether as a by-product of routine data analysis, or as the result of a field visit, must be reported immediately, whatever the normal schedule. To wait until the full survey report is ready is erroneous and to bury significant findings among routine material is unforgivable.

The approach to information transmittal should therefore, be one of willingness to communicate provisional and approximate findings if they are significant. Unfortunately, those fearful of making a mistake tend to err on the side of caution. Careful checking of surprising facts, deliberate appraisal, and avoidance of jumping to conclusions are all valuable attributes. But some willingness to make the best judgement possible on the basis of whatever evidence is currently available is also necessary.

Evaluation, which is a more deliberate process, can work to a less demanding time schedule. Greater depth of analysis and a more thorough presentation will be required. Nevertheless, events may occur that require the evaluation effort to be advanced or accelerated in order to appraise likely consequences, including the need for modified implementation procedures.

Project information can be transmitted in a number of ways:

(a) verbal, informal communication;
(b) regular updating of charts or graphs on the office wall;
(c) formal presentation at project meetings;
(d) memoranda detailing important and urgent findings;
(e) regular reports of a briefing or summary nature;

(f) formal reports of survey results;

(g) analytical reviews of the project data base; and

(h) specific problem analysis with recommendations.

The more formal reporting will be considered in the next section. What, however, is often neglected are the verbal and graphical modes of communication.

Regular updating of a graph on the manager's office wall, showing deliveries of inputs against targets, may be seen by the manager as a significant contribution of the monitoring system. Passing computer printouts through the internal mail is no substitute for a half-hour opportunity to meet with the manager and guide him through the printouts, drawing attention to the highlights. Not only is this useful in itself, it also serves to familiarize the manager with unaccustomed data configurations so that, in time, easy reference is made, even without a personal presentation. The possibility of developing this kind of relationship depends on the integration of the monitoring team into the management structure.

10.3 Reporting

Monitoring and evaluation reports of various types will be needed. The required format of a report depends not only on the type of user for whom it is written, but also the source of the information to be conveyed.

Reports of field trips need to be prepared immediately on return, covering such items as the itinerary, meetings held, observations made, and general impressions gained, together with recommendations for appropriate action.

Results from studies or surveys need to be embodied in the report, even though faster methods may have been used to communicate pressing, significant findings. Such reports must describe the issues investigated, the methodology of the study, the tabulated results, and the findings.

Collection of time-series data for evaluation purposes may have no formal end-of-survey timepoint. Analyses of these series will, however, need to be conducted periodically and the emerging trends reported in the form of updating previous analyses. Comparisons over time will be the focus of such reports and the analyses need to be conducted in some depth and with statistical rigor.

Finally, there are the regular periodic "state of the project" reports, bringing together all relevant materials from the various sources described in Part 5, ranging from administrative records to the major beneficiary surveys. These reports concentrate on presenting an overall view of project progress and provide the means for decisions on adjustments or alterations in project targets and implementation procedures. The periodicity of these reports, e.g., semi-annual or annual, is a matter for decision at the beginning of the project. It should not be so frequent as to absorb an excessive amount of the senior staff's time, thus endangering the maintenance of the quick, regular communication

135

of items of current importance, as described in the previous section.

In view of this wide range of reporting needs, it is not possible to set out a recommended report layout. However, certain guidelines can be offered that are of general validity.

(a) Presentation of Data

Different users require different degrees of detail in the data included in a report. Summary tables and graphs will meet the requirements of many, but others will need to refer to the less aggregated and unrounded data in order to carry out further analyses. Within a project this is not a major problem; the report should provide a summary, with a reference to the source data files or computer printouts to which the project staff have access. For higher authorities, the summary will almost certainly be sufficient; although, here too, the potential for access to the files remains. It should be noted that access to the master files by users other than those immediately responsible should be according to an agreed clearance and approval system. Issues of confidentiality of the data may arise, and if the files are maintained on computer discs or tapes, the security of these needs to be protected in the sense that unauthorized changes to the records must not be allowed.

As well as providing summary tables, the report will contain summary measures of the distributions of the data in the tables. Introductory comments on certain of these measures have been given in Part 9. The writer of the report must bear in mind that many users of the data will not be aware that misleading conclusions can be drawn from even simple parameters. A simple arithmetic mean of a highly skewed distribution may lead to erroneous deductions by the user if his attention is not drawn to the skewness, or if an alternative distribution-free measure, such as the median, is not used in conjunction with it.

Consider, for example, the following incomes as measured before and after an intervention (only ten results are shown for convenience but the point made is valid for any size of data file):

Table 14: Income per Beneficiary

	Income	
	Before	After
Beneficiary 1	20	25
2	50	75
3	30	25
4	40	40
5	35	30
6	25	30
7	80	140
8	30	25
9	60	100
10	30	40
Total:	400	530
Mean:	40	53
Median:	30	30

The mean has increased sharply due to the large increase achieved by those beneficiaries (Nos. 2, 7, and 9) who were already the richest; the median is unchanged.

The use of percentages may also confuse, unless carefully used. It is common practice to show individual cell percentages in a two-way table so that the summation to 100 is only achieved by adding both across rows and columns. Most users are unable to glean much from this as there is an automatic tendency to expect percentages to sum to 100 in one direction only, i.e., either vertically or horizontally. If this option is chosen, the choice of direction may be important. Take a simplistic example of a sample of 400 farms in three areas, classified by the number of parcels of land making up the holding.

Table 15: Number of Farms by Number of Land Parcels

	Location			
No. Parcels	A	B	C	Total
1	100	20	16	136
2	60	60	24	144
3	40	40	40	120
Total	200	120	80	400

Little may be detected from a casual examination of this, except that there are more farmers in Area A. Consider, however, Table 15, expressing the same data in percentage form for each area.

No. Parcels	Location		
	A	B	C
1	50	17	20
2	30	50	30
3	20	33	50
Total	100	100	100

This clearly reveals that farmers in the three areas have very different distributions, in terms of fragmentation of holdings, with the modal number being 1, 2 and 3, respectively. On the other hand, to run the percentages across the rows would reveal nothing, except possibly the larger size of Area A.

Simplistic though these examples are, there have been many real cases where a wrong choice of the direction for calculating percentages masked the story that the figures had to tell.

If grouped frequency tables are used, it is advisable to keep the width of each class interval the same, with the exception of an open-ended highest value class. Varying class intervals may lead the user to draw wrong conclusions about modality or the skewness in the distribution.

(b) Attribution of Accuracy and Significance

Whether the data presented originate from records, case studies, or sample surveys, the presenter should discuss the likely accuracy of the data and indicate which findings are significant in view of the margins of error involved.

The analysis of administrative records may present little difficulty if based on an examination of the total file which supposedly includes all relevant cases. The quality of the record-keeping may, however, be variable and such variations should be noted.

Case studies, as discussed in Part 6, present the reporter with a difficult decision regarding the general validity of the findings based on a very few cases. As discussed, it will not usually be possible to make inferences from a limited number of cases (perhaps purposively selected), to the population at large. This is certainly true for numeric estimates of totals, ratios, etc. However, a case study will have the objective of studying the internal relationships of the variables that cause a particular phenomenon to occur. If a consistent pattern emerges, a plausible inference may be made as to the causality of the phenomenon, which, it may be suggested, is likely to be repeated in other cases where the circumstances are similar. That is to say, inference of casuality and relationships may be made in a general way on the basis of plausibility, as long as a suitable cautionary note is struck. For example, a case study may examine in depth a few farmers who adopted a certain project recommendation and a few who did not. Assume it is found that the reluctance to adopt the new pattern of farming was due quite logically to the perception of a labor constraint at a particular period when the new practice makes heavy demands. Assume also that the adopters have more family labor available than the nonadopters and did not perceive the same constraint. The conclusion that adoption/nonadoption is affected by the availability of family labor may be a plausible inference for the general population in the area, even though the numbers affected by limited labor availability cannot be calculated from the study.

The communication of such findings requires skill in interpretation and presentation. The temptation to claim too much must be resisted, particularly, when the pattern is inconsistent and blurred by a range of extraneous factors. It may be a worthwhile conclusion that no discernible pattern exists —at least one set of hypotheses can be rejected, avoiding concentration by management on the removal of an irrelevant constraint.

Results from sample surveys, properly designed and executed, enable numerical inferences to be made to the population at large with calculable margins of sampling error attached. The procedures for doing so were introduced in Part 7, although from the different perspective of calculating the required sample size for a given sampling error. The formulae for calculating the error margin for any sample design are to be found in standard texts on sample surveys.

Having calculated the sampling error, the question arises of determining which estimates are significantly different, either from one another, or from a standard of comparison, so that conclusions may be drawn with reasonable confidence. Again, the techniques are to be found in the standard texts. What is emphasized here, as in Part 7, is the danger that the reporter may set too demanding a standard before describing a result as significant. The use of the "classic" 95% confidence level implies that any statement of a significant change or difference has a very high chance (19:1) of being correct. Without a recommendation emerging from the survey, an existing procedure may be maintained when there is, let us say, a 3:1 chance that a change would be beneficial. Or, if a decision must be made, the manager may be forced to rely on subjective judgments unaware that the data showed a strong, but less than certain support, for one particular course.

The relative costs of making a change when it should not have been made, as compared to leaving unchanged a procedure that required changing, are the determinants in deciding the level of confidence to be attached to significance testing. In one case, the cost of making a change may be very high. Considerable confidence in the evidence for making such a change will reasonably be demanded. In another, little may be lost in making a proceduralchange, whereas, to let matters "drift" may amount to bad management. The decision to change in this case does not require full confidence in the significance of the evidence for it.

Often the implications of the survey data are not perceived in such a major decision-making role. In simpler terms, the user may look at a table of figures, wanting an answer to the question: "Am I on the right track; are yields rising?" An answer: "probably, yes", implying a better than even chance that it is so, may serve a more useful purpose than a conservative, "not known", based on the use of rigorous confidence standards. Such an answer must, of course, be based, not only on the actual increase, but also after consideration of past variability.

Sampling errors can be calculated; nonsampling errors often cannot. Yet, as discussed in Part 7, these may be crucial in any consideration of the accuracy of the estimates presented. The author of the tables should have an intimate knowledge of the source, methods used in collection, problems in analysis, etc. It is a responsibility, therefore, to provide an assessment of the likely biases and observational errors that may affect the results, even though such an assessment lacks the mathematical support that the sampling error calculation commands. To state that the estimate of output is X with a standard error of, say, 4 percent, leaving unstated the suspicion that the measurements were seriously biased, is to portray a totally spurious impression of accuracy. A description of the survey procedures, design, methodology, nonresponse rate, practical problems, results of field checks, etc. may serve the purpose. But, as most readers may skip such introductory sections, the reporter needs to comment on the implications of these in the text accompanying the tables.

(c) The Use of Advanced Statistical Techniques

Some of the tools available to the analyst have been introduced briefly in Part 9. Many of the techniques available are powerful, but the results are easily misunderstood by the nonprofessional statistician, into which category most readers will fall. What is necessary is to communicate the significant findings in an intelligible manner, helping the reader through the technical details. This brings the discussion to the question of the style of the text in survey reports.

(d) The Text

The text accompanying the numerical data presented in a report is intended to supplement the tables and thus assist in the process of converting the data into information. The method of presentation of this essential supplement and the style in which it is written will vary according to the writer; there is no harm in, and much in favor of, such individuality, as long as the ultimate objective of conveying the information clearly to the intended user is achieved.

A report should contain the following information:

(a) the topics under discussion;
(b) the source and reference period of the material presented;
(c) the design and execution of the survey (if a survey was involved);
(d) a summary of the data;
(e) the necessary comments regarding the accuracy of the data;
(f) a review of the highlights and implications of the data; and
(g) where appropriate, the drawing up of options for decision-making.

The order in which these sections are arranged needs careful consideration. Sections (f) and (g) may be given as a summary at the very beginning of the report—taking maximum advantage of the reader's limited timespan of concentration on the particular report. If the findings are important enough, the further attention of the reader to the body of the report may be secured. The background description contained in (b) and (c) may be relegated to appendices or retained as the necessary introduction at the beginning of the main body of the report. Within the report itself (d) may be a very succinct presentation with the more detailed tables also appearing as appendices.

It is, however, important to accompany each summary table with a paragraph or two of text, drawing attention to the facts of significance, providing indications of their likely reliability, and setting them in their correct perspective in the context of the information contained in other tables. The alternative is to consolidate the text in one section and the tables in another. This is usually the less desirable alternative, for two reasons. First, the general reader may not refer to the tables when reading the text, so that the impressions gained are divorced from the underpinning provided by the actual figures. Second, many *users* of the report may turn directly to the tables in order to obtain a figure or set of figures. The text, with its insights and caveats will not be referred to, leading to a danger that the data will be misinterpreted. If the text and tables are integrated, these dangers are, at least, minimized.

The text in reports of this kind should be short and spare in style. The arguments which the report sets out should have a focus and a sharpness in their presentation. Side issues should be treated separately and not allowed to confuse the development of the main thesis. Brevity, however, does not release the writer from the need to indicate his assessment of the evidence; rather, the briefer and more aggregated the presentation, the more important becomes the role of the writer in guiding the reader to the correct interpretation. Highly aggregated summary tables cannot be further examined or investigated by the reader; it is important, therefore, that the writer provides the necessary explanatory comments.

The user of a report, whether it is short or long, is asked to take certain things on trust, namely, that the reporter has been objective and frank in the way the facts are marshalled. It may be assumed that deliberate falsification has not been introduced, but selective omission of contradictory evidence, removal of data from their limited context in order to generalize, and failure to report the possibility of biases may produce, indeed, a false picture.

The monitoring and evaluation officer may judge his success in communicating useful information by the extent to which he is accepted as a vitaland integral part of the management team.

INDEX

A

Administrative records, 18,36,51,
     55,71,82-84,89,138
Adoption rates, 5,10,16,36-39,
     53,55,139
Amenities, access to, 43-45
Anthropometric indicators, 35,
     43-45,55,92

B

Baseline survey, 2,8-10,27-29,47
Beneficiaries, 2,3,5-10,14,18,
     36-39,46,51,89,93
Bias, 31-34,41,71,79-81,
     84,87,88,91,92

C

Case studies, 10,28,29,40,41,43,
     51-58,62-68,72,92,138,139,
     or sample surveys, 53-56
Causality, 3,7,20-23,27-29,52,58,
     63,109,127,128,138,139
Chi-squared, 125,126
Computers
   use of, 15,87,109-112,131,132
Confidence limits, 28,75,76,
     113,128,139,140
Correlation, 78,119,
     123-125,127-131
   intra-class, 85,86
Credit, 36-39
Crop
   production, 3,33-35,
     39-41,79,92,93
   yields, 32-35,40,
     41,76,79,80,91,92
   prices, 33,42,56,91-93
   area, 33-35,40,41,50-65,91
   marketing, 33-35,41,42,91-93

D

Data analysis, 4-8,10,14,18,26-29
     35,39,47,50,78,89,90
     109-131,135
Data collection, 4-9,25-30,34-36,
     47-49,57-59,74,89-111
Data presentation, 113-142
Data processing, 36,109-131
Data sources, 47-56
Demographic & vital statistics,
     43-46
Direct observation methods, 57-68,
     90-93

E

Effects, 3-5,7,32-58
Enumerators, training of, 45,
     102-107
Evaluation, 1-10,20-29,134
   and research, 65
   indicators for   see Indicators
Experimental model, 20-29,52

F

Farm management studies, 37,41,
     52,54,55

G

Gini coefficient   see Inequality
Guidelines for Design of M&E, 1

H

Health, indicators of, 43-45,
     55,91,92
Hierarchical community structure,
     57-59
Holding, 45,50
Household, 41-43,45,46,50,66,70,71

143

INDEX

I

Impact, 3-7,24-27
Income & expenditure, 3,7,31-33,
  42,43,94
Indicators, 9,10,16,26,29-46,
  66,67,79,89-92
  for monitoring, 35-39
  for evaluation, 39-45
Inequality, 119-122
Inputs, 2-6,32,36,58
Interviews with survey respondents,
  54,55,61,62,78,89,90,93-98

L

Lorenz curve   see Inequality

M

Mean square error, 79-81
Memory, 34,40,41,59,93,94
Modus operandi method, 20,21,28
Monitoring, 3-6,10-19, 35-39
  indicators for   see Indicators
Monitoring & evaluation, 1-5,8-13,
  29,30,33,45-50

N

Non-sampling errors, 69,79-81,
  91-93,128-131
Nutritional status   see
  Anthropometric indicators

O

Odds ratio, 123,124
Outputs, 3-7,14,32,39-41

P

Percentages, 137,138
Project design and implementation,
  1-11,18,47
Project management, information
  needs, 1-8,11-19,50,132-142
Project objectives, 2-4,6,7,13,14,
  17-20,30,31,36

Q

Quality of life, 3,6,43-45,70
Quasi-experimental designs, 20,
  25-29,52
Questionnaires, 8,9,53,54,90,94-111

R

Rapid assessment, 34,51,52,
  55,59-62,72
Recall period, 9,40,43,59,93-95
Reference period, 40-43,93-95
Regression analysis, 124-131
Reporting, 12,18,36,64,132-142
Response, 2,6,7,89,90
  errors   see Non-sampling errors

S

Sampling
  general, 69-88
  probability, 8,52,70-74
  purposive, 8,9,52,72-74
  quota, 9,72-74
  simple random, 77,81,86
  cluster & multi-stage, 71,78,
    81,84-86
  stratification, 81-86